Teaching Fractions through Situations:
A Fundamental Experiment

Guy Brousseau • Nadine Brousseau
Virginia Warfield

Teaching Fractions through Situations: A Fundamental Experiment

 Springer

Guy Brousseau
Université de Bordeaux
Bordeaux, France

Nadine Brousseau
Bordeaux, France

Virginia Warfield
University of Washington
Seattle, WA, USA

ISBN 978-94-017-8420-7   ISBN 978-94-007-2715-1 (eBook)
DOI 10.1007/978-94-007-2715-1
Springer Dordrecht Heidelberg New York London

Preface

We would like to thank James King, Virginia Stimpson and Marion Walter for their very helpful comments on this book and for the encouragement that came with those comments.

We would also like to thank all of the teachers at the École Michelet in Talence for their support, encouragement and very hard work without which the project we are describing could never have happened.

Readers who would like to find out more about the École Michelet or about the field of research whose foundation this book describes can do so at http://faculty.washington.edu/warfield/guy-brousseau.com/.

Contents

1 Why These Adventures? 1

  A Few Words by the Anglophone Author 3

    First an Introduction to All Three Authors 3

    Next the Background of the Teaching Project Itself:

    How and Why It Came to Exist 4

  Introductory Remarks by Guy Brousseau 7

2 The Adventure as Experienced by the Students 9

  Module 1: Introducing Rational Numbers as Measurements 10

    Lesson 1: Measurement of the Thicknesses of Sheets

    of Paper by Commensuration 10

    Lesson 2: Comparison of Thicknesses and Equivalent

    Pairs (Summary of Lesson) 15

    Lesson 3: Equivalence Classes – Rational Numbers

    (Summary of Lessons) 16

  Module 2: Operations on Rational Numbers as Thicknesses 18

    Lesson 1: The Sum of Thicknesses (Summary of Lesson) 18

    Lesson 2: Practicing the Sum of Thickness. What Should

    We Know Now? 19

    Lesson 3: The Difference of Two Thicknesses (As Measure) 21

    Lesson 4: The Thickness of a Piece of Cardboard Composed

    of Many Identical Sheets: Product of a Rational Number

    and a Whole Number 22

    Lesson 5: Calculation of the Thickness of One Sheet:

    Division of a Rational Number by a Whole Number 24

    Lesson 6: Assessment 25

Module 3: Measuring Other Quantities: Weight, Volume and Length 26
   Lesson 1: Making Measurements 26
   Lesson 2: Construction of Fractional Lengths:
   A New Method Appears 28
   Lesson 3: Comparison of Methods, and Demonstration
   of Equivalence... 30
   Lesson 4: Fractions of Collections... 31
Module 4: Groundwork for Introducing Decimal Numbers 33
   Lesson 4: Whole Number Intervals Around a Fraction 33
Module 5: Construction of the Decimal Numbers..................................... 36
   Lesson 1: Bracketing a Rational Number with Rational Numbers:
   Chopping up an Interval... 36
   Lesson 2: Bracketing a Rational Number Between Rational Numbers,
   Shrinking the Intervals, and Observing Decimal Filters........................ 39
   Lesson 3: Representation on the Rational Number Line 41
   Lesson 4: From Writing Decimal Rational Numbers
   as Fractions to Writing Them as Decimal Numerals.............................. 45
Module 6: Operations with Decimal Numbers (Summary)........................ 48
Module 7: Brackets and Approximations (Summary)................................ 49
Module 8: Similarity 51
   Lesson 1: Enlargement of a Puzzle 51
   Lesson 2: The Image of a Whole Number.. 53
   Lesson 3: The Image of a Fraction 55
   Lesson 4: The Image of a Decimal Number 60
   Lesson 5: Division of a Decimal Number
   by 10, 100, 1,000, … (Summary) 62
Module 9: Linear Mappings... 63
   Lesson 1: Another Representation of the Optimist
   (Lesson Summarized) 63
   Lesson 2: (Summary of Lesson)... 65
   Lesson 3: Lots of Representations of the Optimist
   (Summary of Lesson).. 65
   Lesson 4: Good Representations, Not So Good Representations 68
   Lesson 5: Change of Model 70
   Lesson 6: Reciprocal Mappings.. 73
Module 10: Multiplication by a Rational Number..................................... 74
   Lesson 1: Multiplication by a Rational Number..................................... 74
   Lesson 2: Multiplying by a Decimal (Summary of Lessons) 78
   Lesson 3: Methods of Solving Linear Problems
   (Summary of Lessons).. 78
   Lesson 4: The Search for Linear Situations
   (Summary of Lessons) 79

Module 11: The Study of Linear Situations in "Everyday Life" 81
   Lesson 1: Fraction of a Magnitude 81
   Summary of the Remaining Paragraphs and Sections of Module 11 87
Module 12: More on the Problem Statement Contest................................ 90
   Lesson 1 90
Module 13: New Division Problems in the Rationals............................... 93
   Lesson 2: (Extract) Division as Reciprocal Mapping
   of Multiplication (The Term Is Not Taught to the Students) 94
Module 14: Composition of Linear Mappings... 101
   Lesson 1: The Pantograph.. 101
   Lesson 2: Composition of Mappings: First Session 103
   Lesson 3: Composition of Linear Mappings: Designation
   of Composed Mappings 107
   Lesson 4: Different Ways of Writing the Same Mapping...................... 110
   Lesson 5: Rational Linear Mappings 114
Module 15: Decomposition of Rational Mappings. Identification
of Rational Numbers and Rational Linear Mappings 118
   Lesson 1: Decomposition of Rational Mappings.................................. 118
   Lesson 2: The Meaning of "Division by a Fraction"
   (Summary of Lessons) 121
   Lesson 3: Division of Decimals... 124

3   The Adventure as Experienced by the Teachers 127
Background of the Project 127
The Relationship with the Theory of Situations 131
The Perspective of the Teacher 134
   Observable Aspects of *Connaissances* 140
   Manifestation of *Savoirs* 140
   What Then Are the Causes of Learning and the Reasons
   for Knowing? 144
How Does the Teacher Use Assessment of and Within the Curriculum?... 145
   The Assessment of Students and Groups of Students............................ 146
The Types of Situations That Appear in the Lessons................................ 147
   The Types of Didactical Situation and How They Are Conducted........ 148
   The Types of A-didactical Situation and How They Are Conducted 150
   Presentation of the Rules of the Game... 153
   Evaluation in A-didactical Situations 154
Obsolescence... 155
Isolated Evaluation of *Savoirs* and Constant Evaluation;
the Necessity of the Uncertain and the Implicit....................................... 156
The Play of the Real and the Fictional.. 156
The Inexpressible, the Said and the Unsaid 157
Further Aspects of the Teachers' Adventures 157

The Mathematical Organization of the Curriculum 159
Mathematical Commentary on Chap. 2 160
    The Temporary Replacement of Fractions by Commensurations 161

4 **The Adventure as Experienced by the Researchers** 165
Warfield Introduction Concerning the History of and Voice
in Chapter 4 165
Brousseau Introduction to Chapter 4 166
Prelude (1960–1970) 170
    The Sources 170
    The Adventure of "Modern Mathematics" 171
    The Subject of the Studies Proposed by Lichnerowicz 172
    The Background of the Future Research 173
    Experimentation: How and in What Form? 175
    Our Experiments 177
    From Experiments to Theories: And a Science? 178
    The Framework 179
    Observation 181
    Reflections on This Ambitious Project 181
The Foundations (1970–1975) 182
    The IREM [Instituts de Recherches pour l'Enseignement des
    Mathématiques]; the Bordeaux IREM 182
    The COREM (1973) 186
Further Developments over Time 186
    The Diplôme d'Études Avancées de Didactique
    des Mathématiques (DEA) 1975 186
    The Doctorate of Didactique of Mathematics,
    Part of Mathematical Sciences 187
    Documentation 188
    Research Organizations in Didactique of Mathematics 188
    The Current State of Didactique of Mathematics 189
Further Commentary on Professor Lichnerowicz's Challenge 189
    Institutional Difficulties 189
    Difficulties in Experimentation 190
    Possibilities for Experiments 190
    A New Conception of What It Means to Teach 191
    Conception of Teaching 191
    Construction of Alternatives 192
    The Contributions of Piaget 192
    The Notion of Situation 193
    First Questions 194
    A Child and a Concept 195
    Models of a Genesis 195
    The Standard Presentation of Mathematical Concepts 196

Connaissances and *Savoirs*.. 197
The Place for *Connaissances*: The Types of Mathematical
Situations and Theories.. 198

5   **Expansions and Clarifications** 201
   Connaissances and *Savoirs*.. 201
   Didactical Situations 202
   Institutionalization 203
   Didactical Contract 204
   Connaissances and Epistemological Obstacles 205
   Metadidactical Slippage... 206
      Various Examples of Uncontrolled Chains
      of Metadidactical Slippages.. 208
      The Slippages Studied in This Curriculum... 208
   Evaluations... 209

Bibliography 211

Index... 213

Acronyms

ARDM	*Association pour la Recherche en Didactique des Mathématiques*
	Association for Research in *Didactique* of Mathematics
CIEAEM	*Commission Internationale d'Études et d'Amélioration de l'Enseignement des Mathematiques*
	The International Commission of Studies and Improvements of the Teaching of Mathematics
COREM	*Centre d'Observations et de Recherches sur l'Enseignement des Mathématiques*
	Center for Observation and Research on Mathematics Teaching
CNRS	*Centre National de la Recherche Scientifique*
	National Center of Scientific Research
CRDM-Guy Brousseau	Centro de Recursos de Didáctica de las Matemáticas - Guy Brousseau de l'IMAC
	IMAC Resource Center for Guy Brousseau-*Didactique* of Mathematics
CRDP	*Centre Régionale de Documentation Pédagogique*
	Regional Center of Pedagogical Documentation
CREM	*Centre de Recherches pour l'Enseignement des Mathématiques*
	Center of Research for the Teaching of Mathematics
DEA	*Diplôme d'Études Avancés*
	Diploma of Advanced Studies
ICMI	The International Commission on Mathematical Instruction
IMAC	Institut Universitari de Matemàtiques de Castellon
	Mathematics Institute of Castellon
IREM	*Institut de Recherches pour l'Enseignement des Mathématiques*
	Research Institutes for Mathematics Teaching
ViSA	*Vidéos de Situations d'enseignement et d'Apprentissage*
	Videos of Teaching and Learning Situations (see http://visa.inrp.fr/visa)

Chapter 1
Why These Adventures?

This book is intended to be read by teachers, researchers in education, mathematicians, and anyone else who is curious about what educational research has to say about the teaching of mathematics. It centers around a set of lessons on rational and decimal numbers. The lessons came into existence to validate the Theory of Situations, a basic tenet of which is that children can best learn a mathematical concept by being put into a very carefully designed situation where achieving some goal requires them to invent or discover the concept, and their prior knowledge enables them to do so.

The core of the book is the day-by-day journal of a fifth grade class in which the teacher reports every stage of what she presented in 65 lessons on rational and decimal numbers, and what happened with it. The journal was originally produced to enable two parallel classes to reproduce the lessons. The lesson sequence was conceived in 1972–1973 and was considered stable by 1975–1976. Enriched by various observations made by succeeding teachers, the sequence was officially reproduced every year in two parallel classes until 1999.

This lesson sequence is one of a number realized in the COREM (*Centre d'Observations et de Recherches sur l'Enseignement des Mathématiques*), a school set up specifically for observation supporting mathematics education research. A description of the school and its functioning can be found in Chap. 3, while Chap. 4 provides the origins of its conception as a research necessity. The lessons carried out there played a central role in the development of *Didactique* – a program of scientific research in mathematics education whose structure is unique to France, but whose contributions are valid and valuable everywhere that mathematics is taught.

Untangling the web of ideas, experiments, discoveries, hypotheses and proofs involved in a new teaching project is a long, perilous and debatable task. What we have to tell is thus the tale of three adventures.

One is the adventure of researchers opening up a new territory. It is certainly interesting, but it is complex and breaks with too many concepts and venerable habits of thought to be easily accepted without the support of the observations that

G. Brousseau et al., *Teaching Fractions through Situations: A Fundamental Experiment*,
DOI 10.1007/978-94-007-2715-1_1, © Springer Science+Business Media B.V. 2014

provide its experimental foundations. Some of the many research results produced in a variety of fields:

- Elementary school students are able to construct, understand and practice fundamental mathematical concepts, using the modern mathematical and epistemological organization of those concepts.
- Factors that cannot be formally and directly evaluated, such as things left unsaid and knowledge that cannot be expressed or has not been decided, play an essential role in the elaboration, manifestation, learning and teaching of recognizable knowledge. Methods that eliminate the action of such factors are less effective: when they are used, only the students who are capable of filling in on their own what was left unsaid in the texts that are taught can make progress.
- Radical constructivism does not work as a general model: Institutionalization is indispensable.

The adventure of the researchers constitutes Chap. 4.

The second adventure is that of the teachers, recounted in Chap. 3. It is also captivating. They threw themselves into a scientific episode that was fascinating for them, but strange to them and very much of a disruption of their standard work as teachers. To appreciate that adventure, the reader needs to bear in mind that the teachers whose actions are being directed and recorded in the tale that follows were stepping from a familiar and comfortable terrain into a completely new teaching world in which many of the familiar landmarks had been removed or disguised.

But above all we are eager to introduce the reader to the adventures of the students as they took part in the classes, recounted in Chap. 2. What were the conditions in which they produced and learned some difficult mathematics? In what ways were their mathematical activities closest to the activities of mathematicians? To know that we had to make the conditions of their work explicit, with precision, as they were planned and as they were realized by the teachers. Likewise we had to make their reactions clear – the significant ones that made it possible to pursue the process. The "didactical files" that we have translated provide the best mechanism for following the adventure step by step from the point of view of the students. They were established in order for teachers – the original ones or their successors – to be able to reproduce the lessons. They were reproduced at least 50 times with completely similar results.

The texts that we present or describe in Chap. 2 were thus carefully designed to enable the lessons exactly as described to be reproduced in their original context. On the other hand, they were not designed for the lessons to be exported. They were carried out in an institution that was specifically created to permit this kind of experiment to be carried out in conditions that were secure for the teachers and for the students. This is especially true in that the options we chose were not those that we would recommend for development. They absolutely do not prefigure a curriculum to be developed in ordinary classes. Their sole objective was to provide scientific answers to some essential questions.

If this does not provide a model that can transfer directly into the day-to-day life of a teacher, what does it offer? It offers encouragement and hope, by directly demonstrating forms of teaching and learning that mathematics educators and philosophers have been trying for two centuries to promote. We hope it will encourage all the people who aspire to improve teaching itself, which is currently suffering seriously from the increasing divergence between society's requirements for education and its obstinate refusal to call into question obsolete ideologies and inappropriate scientific practices. Yes, students can learn mathematics, and learn it well, by taking part in mathematical activity. Not only that, but they can thoroughly enjoy doing so. What else is at the heart of all of our endeavors?

A Few Words by the Anglophone Author

The content of this book is completely international. The activities of the children, the decisions of the teachers and the explorations of the researchers are part of a fabric of mathematics education that increasingly is spreading worldwide. However, a certain amount of the background for the teaching project that is central to the book is unfamiliar to most readers outside of France, and knowing the background of the book itself may help enrich the reading of it, so as a lead-in to a book that is very much a joint effort we will present a few paragraphs that are specifically a Warfield production.

First an Introduction to All Three Authors

Guy Brousseau has had a long and notable career in mathematics education research, for which the most telling evidence is probably his having been awarded the first Felix Klein Award from the International Commission on Mathematics Instruction, in recognition of "the essential contribution Guy Brousseau has given to the development of mathematics education as a scientific field of research, through his theoretical and experimental work over four decades, and [of] the sustained effort he has made throughout his professional life to apply the fruits of his research to the mathematics education of both students and teachers."[1] His background, determination and reflections, combined with some favorable circumstances, led him to conceive of, create and sustain both a wide-ranging program of coherent, flexible and scientifically based research and the necessary institutions, including a school, to carry out and develop that research. The program has been successful thanks to the help of numerous collaborators whom Brousseau managed to interest in his projects, and to the encouragement and support that he was given. In particular, it was at the school he helped create that the curriculum here described was taught for many years, starting in the early seventies.

Nadine Brousseau's career was in elementary school teaching, and she was among the initial teachers in the research school. This was ideal for two reasons: she was able to confer with her husband long and deeply about the intentions and plans for the lessons, and the results and implications of what happened when she taught

---

[1]http://www.mathunion.org/icmi/other-activities/awards/past-recipients/the-felix-klein-medal-for-2003/

them. Her contribution was irreplaceable and decisive. In addition, she kept extremely good records, both of the proposed lessons (including the elements added when the two Brousseau's continued their discussions long after their fellow researchers and teachers had gone home) and of the class response to them. Her notes became the functional memory of the project, and her present memories enhance and enrich the recorded ones.

This author (Virginia Warfield) came onto the scene considerably later. In the course of a career that combined mathematics and interesting ways to teach it at both elementary and university levels, I had become increasingly interested in mathematics education as a field. A fortunate sequence of events led me to the work of Guy Brousseau and to the discovery that it was very little known in the English speaking world. My first work was with Nicolas Balacheff who, with translating and co-editing by Martin Cooper, Rosamund Sutherland and myself, published Brousseau's *Theory of Didactical Situations in Mathematics* (Brousseau, 1997).

My work on that book resulted in a partnership with Brousseau himself from which so far a number of articles and talks have emerged, as well as a small introductory book. Four of the articles were a series in the Journal of Mathematical Behavior (Brousseau, Brousseau, & Warfield, 2004, 2007, 2008, 2009), covering separate parts of the Rational and Decimal Number curriculum under discussion here. Eventually we decided that the articles needed to be assembled and expanded into a book.

As should be clear, this thoroughly asymmetrical set of positions leads to some variation in the meaning of the word "we". Since, on the other hand, the variation produces no ambiguities, we (in this case all three authors) have decided to leave it.

Next the Background of the Teaching Project Itself: How and Why It Came to Exist

Part of that background begins in the 1960s, when a substantial international group of mathematics education researchers agreed to the need for more serious, coordinated, collaborative research. In France, part of the response to this need was the establishment of a number of IREM's – Research Institutes for Mathematics Teaching. Guy Brousseau was an enthusiastic supporter of this development, and was instrumental in bringing a very early IREM to the University of Bordeaux, where he was on the faculty. He felt, though, that although an IREM was necessary, it was not sufficient for the level of scientific focus he envisioned. To achieve that level, he spent a lot of time and a huge amount of energy which jointly paid off in the creation of the COREM (Center for Observation and Research on Mathematics Teaching). This center took the form of a school, the École Michelet, which was a regular public school in a blue collar district on the edge of Bordeaux equipped with a carefully constructed set of research arrangements. On the physical side, the arrangements consisted of an observation classroom in which classes would occasionally be held – often enough so that the students found them routine. The classroom was equipped with a multitude of video cameras and enough space for observers to sit unobtrusively. Other arrangements were far more complex, involving an extra teacher at each level and an agreement among the teachers, administrators and researchers setting out the responsibilities and rights of each. Nothing involving that many humans could possibly glide smoothly through the years, but the fundamental idea proved robust, and the École Michelet functioned as a rich resource for researchers for two and a half decades.

Another part of the background has roots that can be traced back through the generations, but came to the foreground in the 1960s under the title of constructivism. The title stems from the underlying tenet that knowledge is constructed in the human mind rather than absorbed by it. Applications of that tenet range from the radical constructivist belief that absolutely no information should be conveyed to students directly, to the naïve conviction that having children manipulate some physical objects that an adult can see to represent a mathematical concept will result in the children understanding the concept itself. Guy Brousseau had studied many of them, but while he found many interesting points, he felt that so far there was a serious lack of solid research in support of the theory itself. With his fellow researchers he therefore set himself the goal of taking some serious piece of mathematics and proving that in certain conditions the children – all the children, together – could create, understand, learn, use and love that mathematics. Accompanying that goal was the goal of studying the conditions themselves.

Clearly the mathematics to be used for this experiment had to be both significant and challenging. After some consideration he made a choice that will resonate with elementary teachers worldwide: fractions, or more properly, rational and decimal numbers. He had, in fact, some reservations about whether rational numbers should be taught at all, but they were firmly part of the national. They had a further virtue: the experimental curricula he had in mind for the very youngest classes introduced them to numbers in such a way as to permit the construction of all the epistemological and mathematical bases of the fundamental numerical structures. Part of the objective was to prepare them for much later studies – reflective, mathematical and formal studies starting at the first year of the secondary level aimed directly at mastering basic symbolic, algebraic and analytic instruments. The study of rational and decimal numbers provided a point of articulation between these two projects.

Having made this choice, he then spent a lot of energy and time doing research into the different mathematical aspects of both the rational numbers and the decimal numbers, as well as possible ways of generating them. He also looked into the history of how each has been taught in different cultures and historical contexts. One of his conclusions was that a major source of learning difficulty is that although rational numbers are used in several very distinct ways – among others as measurement (3/5 cm), as a proportion (this thing is 3/5 as long as that thing), and as an operation (take 3/5 of this quantity) – they are generally taught as if all the meanings were equivalent. The result is that the student must accept many things simply on the basis that the teacher says so, and in the long run has no coherent foundation for the concepts. This conclusion led to the mathematical structure of the curriculum presented here. By way of a roadmap, we will sketch the resulting order here. A more mathematical description will be found in Chap. 3. For a considerably more detailed description of both the background and the decision procedure, see Chap. 4 of the Theory of Didactical Situations in Mathematics (Brousseau, 1997).

The first lessons are taken up entirely with commensuration[2] and its consequences. The children first work with different thicknesses of paper and realize that even though they cannot measure a single sheet, they can distinguish the papers by specifying how many sheets it takes to make up 2 cm, or alternatively how thick 50 sheets are. Deep familiarity with that idea paves the way for developing an understanding of equivalence and the basic operations. That understanding is solidified with some work generalizing the results to measuring weights of nails, volumes of glasses and lengths of carefully selected strips of paper.

The following set of lessons works with decimal numbers. In a series of challenges to find smaller and smaller intervals around some rational number, the class discovers the virtues and some of the working principles of using numbers whose denominator is a power of ten. Once they are secure with that, they begin to use decimal notation for these convenient objects.

With their grasp of rational and decimal numbers as measurements now reasonably solidified, the students then progress to a more active aspect, using them first to enlarge a tangram-like puzzle, then to enlarge and reduce a variety of items. The rest of the curriculum is devoted to deepening mathematical connections, broadening applications and enlivening problem-solving using these concepts.

The remaining element of background concerns the format for the learning adventure itself. Brousseau, in the course of teaching elementary school for several years, reading voraciously and maintaining on-going lively discussions with an array of people that included teachers, university professors, psychologists, linguists, teacher educators, administrators and even a priest had developed his own take on constructivism, which took the form that he eventually called the *Theory of Situations*. His idea was that for children to learn a concept they should be put into a Situation (a very carefully orchestrated classroom situation or sequence of situations) in which in order to resolve some problem or win some game they would need to invent the concept in question. He was strongly committed to this theory, but had an equally strong commitment to the principle that before people were asked to accept it they should be presented with solid research validating it. This pair of commitments helped fuel his drive to create the COREM. Once it was created, his first goal was to design research to test the theory. At the heart of that research was the curriculum that provided the adventure of Chap. 2.

One final note: this curriculum is sufficiently enticing, both mathematically and pedagogically, to give the impression that it should and could be simply picked up and transplanted into other classrooms. This was not Brousseau's intention in producing it, and he warns repeatedly and vigorously against that illusion. It does indeed illustrate a wonderful kind of teaching and learning, and it provides thought-provoking insights and ideas with direct or indirect application to the classroom. On the other hand, the many iterations of successful use of the curriculum itself were all

---

[2]Commensuration is the measurement of things in comparison to each other rather than in terms of a unit.

carried out with the extraordinary support provided by the COREM, and Brousseau feels strongly that an attempt to use it without that support would be likely to have disastrous consequences.

Introductory Remarks by Guy Brousseau

I am very grateful to Virginia Warfield, who has worked hard – and made me work hard – for 20 years to make accessible to the American public the texts of one of our most sophisticated instruments of research. It has the most innocent of appearances as a curriculum – the chronicle of an adventure, programmed down to its details, that the students and their teachers lived and above all that others succeeded in reliving identically. An adventure for the students in the sense that the curriculum gives them the sense of having a lot of space for initiatives, experiments and personal reasoning with goals that seem to them objective and that they are able to believe the teacher does not know … but an adventure also for the teachers who always wonder whether the Situations, even though minutely calibrated and reproduced year after year, will really once again permit them to achieve the desired results: the learning in common of a common mathematical culture shared by all of the students in the class. The cost of the apparent freedom of the students is a no less apparent drastic reduction in the freedom of the teachers.

This curriculum was not made to be used in other classes. The sole purpose of the reproducibility was to consolidate the scientific observations that we needed in order to test certain hypotheses. The lessons had above all the property of making apparent the enormous complexity of the acts of teaching: that of the conception, to be sure, but even more that of the carrying out of the lessons. The fact that teaching is a complex activity and passably mysterious is accepted in theory by our societies – but they don't really know what that means! They absolutely do not take the complexity into account when it comes to studying the work of teaching. They intervene authoritatively in the educational system on the basis of grossly erroneous conceptions. They are not even capable of identifying the specific field of science: the need is to understand a phenomenon and they look only at the actors. The consistency and validity of the concepts in question need to be verified, and instead they look only at their use and market value!

The COREM that we called our "Didactron" was a center for anthropological observation: with their consent, we observed as anthropologists the life of a tribe of teachers. Believe me, this is not an easy approach, even for those taking part in it One among the collaborators and teachers of the COREM was my wife Nadine Brousseau née Labesque, who played an important role in all the steps of the project. She helped me as a collaborator to study didactical versions of the Situations, and as a teacher to present them with her colleagues to the pupils in the school Michelet de Talence for 14 years before her retirement. She also helped with the work of redaction of the script prepared in common and the transcription of remarks and observations. She wrote the first stage of our manual "Rationnels et décimaux dans la scolarité obligatoire" (Rationals and Decimals in Basic School) published by the IREM of Bordeaux. This text, which was produced in 1985, was reserved for researchers in *Didactique*. More than 2,000 copies were sold.

Another invaluable collaborator was Denise Nedelec (known as Denise Greslard), who experimented with the curriculum protocol from 1987 to 1999 with great care and dependable success, and made many fine observations.

We thought the teachers would want to eliminate these lessons after 2 or 3 years, as soon as we had sufficiently observed the phenomenon of the obstacle. Among many challenges was the fact that the least interaction with the students obligated the teachers to interpret their declarations, put out in the system of commensurations, by translating them into the teacher's own knowledge system, that of fractions, and then make reciprocal translations to continue the lesson. Knowing that even though the results are the same the proofs are often different in the two systems it is easy to see that the mathematical exercises produced a lot of stress for them and made the role of the culture in mathematical activity palpable, often cruelly so. Our observations in this context largely confirmed what we had seen of the difficulties of students as they pass from one system to the other.

We were therefore extremely surprised at the end of the experiment when the teachers expressed their desire to keep these lessons in the curriculum despite these difficulties. This reaction led us to understand that in certain cases jumps in complexity can be highly effective. The classical approach is to deconstruct material to be learned so as to keep the amount of information delivered by each lesson more or less constant and optimal. Our experiments demonstrated that in certain particular circumstances this rule can be violated to very good effect. Most of the rules of teaching as practiced are only valid in the absence of deeper and more specific knowledge about the conditions of teaching.

I hope that this gives our reader an idea of what we are offering in this work. I ask them to extend us some credit and to search for good questions before searching for answers. Video recordings of some of these 65 lessons, realized in the course of the 25 years of the COREM, are collected at the ViSA site (Vidéos de Situations d'enseignement et d'Apprentissage http://visa.inrp.fr/visa) to which researchers have access. In addition, all the homework and exercises of all the students from 50 realizations of these lessons can be consulted at the University Jaime 1 de Castillon (Spain) which can make copies of them (made anonymous),

Our curriculum presents a wide variety of types of lessons. Each one has its role and its necessity. But there is absolutely no pedagogical, didactical or epistemological message hidden in them – only questions and occasions to reflect and make discoveries yourself. Try! These are not riddles. Sometimes I give my answers. Compare them to your experiences. The curious could, if they like, launch themselves into a study of the Theory of Situations. So if something astonishes you, ask yourself questions, whether it has to do with the conception, with the conduct of the lesson or with the result of the lessons. Ask us your questions, and we will think about them with you.

Chapter 2
The Adventure as Experienced by the Students

Before taking the reader into the classroom, we need to introduce the children who will be found there. Other chapters introduce the school in which the classroom was located and the teachers who carried out the lessons, but here we are focusing on the students in a particular classroom. Who were they? The first key piece of information is that since the school was an essential element of the COREM (Center for Observation and Research on Mathematics Teaching) admissions were emphatically not selective. The school was the public school for a blue collar neighborhood, and its students were the ones who lived around it. Parents were kept informed about the unusual aspects of the teaching, but there were no special requirements or requests of them. On the other hand, the lessons we visit took place in the fifth grade with students most of whom had been at the school since age three or four, so all of their expectations for what would happen in a mathematics class were built around the kind of activity and responsibility we see in action. They needed no persuasion to involve themselves.

Enjoy joining them!

G. Brousseau et al., *Teaching Fractions through Situations: A Fundamental Experiment*,   9
DOI 10.1007/978-94-007-2715-1_2, © Springer Science+Business Media B.V. 2014

Module 1: Introducing Rational Numbers as Measurements

Lesson 1: Measurement of the Thicknesses of Sheets of Paper by Commensuration

The objective of the first set of lessons is to have the students invent a way to measure something so thin that their previous methods of measurement cannot be applied. The challenge is to find the thickness of a sheet of paper, which they clearly can't do directly with the usual measuring devices. They discover that "repeating the thickness" – that is, stacking the sheets of paper – provides sufficient thickness for their rulers to give a reading.

The Set Up

On a table at the front of the classroom are five stacks (or half-boxes) containing 200 sheets each of paper. All the paper is of the same color and format, but each box contains paper of a different thickness from the others (for example, card stock in one, onionskin in another, etc.) The boxes are set up in a random order and labeled A, B, C, D, E. Some of the differences should be impossible to determine by touch alone. The teacher needn't know the exact measurements, since there is no "good measurement" to be discovered.

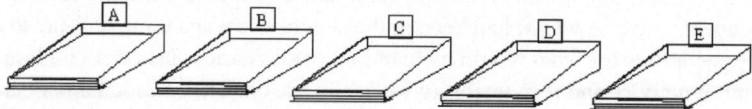

- On another table at the back of the classroom are five more stacks or boxes of the same papers, in a different order, which will be used in phase 2.
- Each group of five students has two slide calipers (a device for measuring thickness, standard in French elementary classrooms)
- The ends of the room are screened from each other in some way – a curtain or a screen.

The Search for a Code

(a) The teacher divides the class into teams of four or five students and presents the situation and their assignment:
   "Look at these sheets of paper that I have set up in the boxes A,B,C,D,E. Within each box all of the sheets have the same thickness, but from one box to another the thickness may vary. Can you feel the differences?"
   Some sheets from each box circulate, so that the students can touch them and compare them.
   "How do businesses distinguish between types?" (weight)

"You are going to try to invent another method to designate and recognize these different types of paper, and to distinguish them entirely by their thickness. You are grouped in teams. Each team must try to find a way of designating the thicknesses of the sheets. As soon as you have found a way, you will try it out in a communication game. You may experiment with the paper and these calipers."

The students almost invariably start by trying to measure a single sheet of paper in order to obtain an immediate solution to the assignment. This results in comments to the effect that "It's way too thin, a sheet has no thickness" or "it's much less than a millimeter" or "you can't measure one sheet!"

At this point there is frequently a moment of disarray or even discouragement for the students. Then they ask the teacher if they can take a bunch of sheets. Very quickly then they make trial measurements with five sheets, ten sheets – until they have a thickness sufficient to be measured with the calipers. Then they set up systems of designation such as:

10 sheets 1 mm

60 sheets 7 mm

or $31 = 2$ mm[1]

In this phase, the instructor intervenes as little as possible. He makes comments only if he observes that the students are not following – or have simply forgotten – the assignment.

The students are allowed to move around, get more paper, change papers, etc.

When most of the groups have found a system of designation (and the children in each group agree to the system or code) or when time runs out, the teacher proceeds to the next phase: the communication game – going on even if not every group has found a system.

The Communication Game

"To test the code you just found, you are going to play a communication game. In the course of the game you will see whether the system you just invented actually permits you to recognize the type of sheet designated. Students on each team are to separate themselves into two groups: one group of transmitters (two students) and one of receivers (two or three students). All the groups of receivers go to one side of the curtain, and the groups of transmitters to the other. The transmitters are to choose one of the types of paper on the original table, which the receivers can't see because of the curtain. They will send to their receivers a message which should permit them to find the type of paper chosen. The receivers should use the boxes of paper set out on the second table at the back of the classroom to find the type of paper chosen by the transmitters.

---

[1] This use of the equal sign is incorrect. The teacher will mention it during the discussion time.

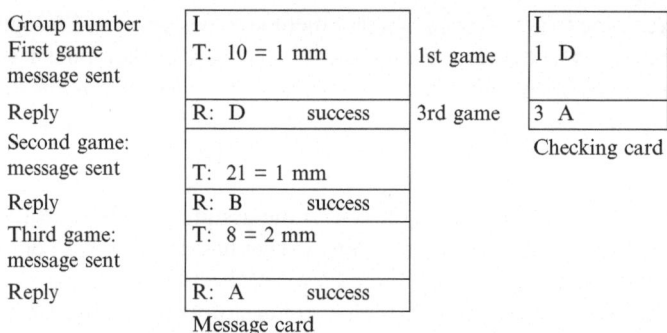

Fig. 2.1 Cards for the communication game

When the receivers have found it and checked it with the transmitters, they become transmitters. Points will be given to the teams whose receivers have correctly found the type of paper chosen by the transmitters."

At the beginning of the game, the teacher puts the curtain in place. Then he

– Passes the messages from the transmitters to the receivers
– Receives the responses of the receivers
– Checks whether this response corresponds to the choice of the transmitters and announces the success or failure to all of the team.

All of the messages are written on the same sheet of paper, which we can call the "message card" (see Fig. 2.1), which the teacher carries back and forth between the transmitters and receivers on the same team, marking whether the receivers have selected the correct paper ("success") or not ("missed"). The team's number is written on the card. In addition, the transmitters write the type of paper that they have chosen on another sheet of paper – the "checking card" – which they keep.

Clearly, the teacher does not introduce superfluous formalism or vocabulary. If certain teams have not arrived at any way of sending effective messages, the teacher could send them back to considering a code together (same assignment as in the first phase). On the other hand, in the first eight identical trials of this material, that never happened. The students always managed to play two or three rounds of the game.

During this game, there are three different strategies commonly observed:

Some choose a particular number of sheets and always measure that number.
Some choose a particular thickness and count how many sheets it takes to make that.
Some look randomly at a thickness and a number of sheets.

The children predictably prefer to choose the sheets of extremes of thickness, either the thinnest or the thickest, to make the job easier for their partners.

Result of the Games and Comparison of the Coding Systems

For this phase, the students go back to their original places in teams of 5, as for the initial phase. The teacher prepares a chart with group names down the side and paper types (A, B, C, D, E) across the top.

Taking turns, each team sends a representative who reads the messages out loud, explains the code chosen and indicates the result of the game. The teacher keeps a record of the groups' messages (and their success) as the reports are made.

The different messages are compared and discussed by the students. Since they are frequently very different, the teacher requests that they choose a common code.

Example: 10; 1 mm
        VT (for Very Thin)
        60; 7 mm

After discussing these, the class chose: 10; 1 mm and 60; 7 mm.

The children rewrite their messages and present them successively in no particular order on the blackboard. Immediately there are spontaneous remarks like "That can't be!" and "That one's OK", etc.

For example: "Group 2 said 30 sheets of paper C were 2 mm thick, but Group 4 said the same number were 3 mm thick. That can't be!"

The teacher announces that if there are disagreements the groups in question should carry out their measurements again.

The session ends with a request to arrange the chosen messages all on the same chart.

Different Types of Inconsistencies

The students' measurements are collected on a chart such as the following (1977)

Type of paper	Group 1	Group 2	Group 3	Group 4
A	19 s; 3 mm	10 s; 2 mm	20 s; 4 mm	
B	19 s; 3 mm		4 s; 1 mm	15 s; 2 mm
C	19 s; 2 mm	30 s; 2 mm	100 s; 8 mm	30 s; 3 mm
				15 s; 1 mm
				20 s; 2 mm
D	19 s; 2 mm		100 s, 9 mm	
E			9 s; 4 mm	13 s; 5 mm
				7 s; 3 mm

Students look for and discuss the inconsistencies. By the end of the session, they have identified categories of errors among the following:

1st category:

If the sheets are of different types, the same number of sheets should not correspond to the same thickness.

Example:

$$19 \text{ s; } 3 \text{ mm} \qquad \text{Type A}$$
$$19 \text{ s; } 3 \text{ mm} \quad - \quad \text{Type B}$$

"That can't be!"

2nd category:

If the sheets are of the same type, the same number of sheets should correspond to the same thickness.

Example:

$$30 \text{ s; } 2 \text{ mm} \quad - \quad \text{Type C}$$
$$30 \text{ s; } 3 \text{ mm} \quad - \quad \text{Type C}$$

"That can't be!"

3rd category:

If there are twice as many sheets of the same type, it should be twice as thick.

Example:

$$30 \text{ s; } 3 \text{ mm} - \quad \text{Type C}$$
$$15 \text{ s; } 1 \text{ mm} - \quad \text{Type C}$$

"That can't be!"

and the students add: "It should be

30 s, 2 mm
        and   because   ×2 | 15 s; 1 mm | ×2
15 s; 1 mm   30 s; 2 mm

4th category:

A difference in the number of sheets shouldn't correspond to the same difference in thickness.

Example:

19 s; 3 mm
                "That doesn't work, because one sheet can't be a
20 s; 4 mm   millimeter thick!"

The teacher makes no explicit reference to the formal use of the concept of pro-portionality, and does not ask it of the students either. On the contrary, she favors the explanations given by the students to whatever extent they are understood, but does not at this stage correct the ones that are not understood.

Didactical Results

At the end of this first sequence, all of the students know within this specific set-up

How to measure the thickness of a certain number of sheets of paper
How to write the corresponding ordered pair
And to reject a type of paper that does not correspond to an ordered pair given to them (if the difference is large enough.)

Most of them are thus able to analyze a chart of measurements to point out inconsistencies making *implicit* use of proportionality.

Those who can't do so seem to understand those who do it.

Order: The children know how to find equivalent pairs. They know how to compare the thicknesses of sheets of paper (many by two different methods).

This knowledge is sufficient to undertake (understand the goal and resolve) the situations that follow.

Lesson 2: Comparison of Thicknesses and Equivalent Pairs (Summary of Lesson)

The first step is a review of the chart produced in the previous lesson. Students first study it silently and make individual observations, then discuss these observations as a class. The chart is corrected either by universal agreement, or, where that agreement doesn't occur, by a re-measurement. This process serves to bring out the idea of augmenting the number of sheets counted in order to distinguish between papers of highly similar thicknesses as well as to exercise further the implicit use of pro-portionality to determine consistency of representations of the same paper.

Working in (non-competitive) groups, students then fill in any empty slots on the chart by counting sheets and then comparing their results with those of other groups. As a confirmation and celebration, they play one more round of the communication game from the previous session, discovering that they are now equipped to handle it even if a couple more types of paper are tossed in. This finishes the second session.

The children must refer with precision to a number of new objects: physical sizes – the thickness of a stack of sheets, the thickness of a single sheet; the *numerical expressions* for these thicknesses: a number of sheets and a number of millimeters for the first, the two numbers combined for the second; some *generic terms* for these denominations: "number", "pair", "ordered pair", etc. This vocabulary is not supposed to be taught with formal lessons. Only the accuracy of the thinking counts. The teacher is faced with the difficult task of helping the use and formulation of these concepts move forward without disturbing the expression of the thought processes. This produces a fragile equilibrium to be maintained and developed.

> **Results** The children know how to adapt the number of sheets chosen to meet the needs of discriminating between their thicknesses (increasing the number if the thicknesses are too close). They know how to find, by calculating, which ordered pairs correspond to the same type of paper. All of them now know how to use proportionality to analyze a chart. Some of them are able to use the relationship of proximity between the pairs. Many of the children have been led to make judgments about statements and to make arguments themselves.

Lesson 3: Equivalence Classes – Rational Numbers (Summary of Lessons)

In the following session the completed chart is once more the center of attention, and the central topics are equivalence and comparison. After getting the students to focus on the chart, the teacher presents some other pairs of numbers and asks which kind of paper each pair represents, then has the students invent other representations, listing all of the accepted ones in the same column on the chart. This provides the occasion for introducing the term "**equivalent**".

"50 s; 4 mm and 100 s; 8 mm are two names, corresponding to different stacks of sheets of the same paper and the thickness of these stacks. We introduce these stacks to identify *the same* object, *the thickness of one sheet*. Since they designate the thickness of the same sheet, the pairs are *equivalent*. 50 s; 4 mm is equivalent to 100 s; 8 mm."

The teacher then produces a new chart with a single name for each kind of paper (the class chooses the name) and the students are told to figure out the order of the papers, from thinnest to thickest. Students work individually, and then discuss their results and their reasoning.

Once an order is agreed on, the teacher introduces another type of paper (fictional this time) and the students figure out where in the ordering it belongs.

As a final step, the teacher returns to the chart with columns containing equivalent ordered pairs for each type of paper and introduces the standard notation a/b to designate the thickness and differentiate it from the varied ways, with a variety of stacks of sheets, they have been using to determine the thickness of one sheet.

The teacher points out that this not only makes it possible to designate the entire class of equivalent pairs, but also gives a designation for the thickness of a single sheet of paper. Thus, a s; b mm designates a stack of sheets and its thickness, b/a mm. is the thickness of each sheet.

The teacher uses the words "ordered pair" and "fraction" without giving a definition for distinguishing the type of notation required. There are many fractions that designate the same thickness.

The lesson finishes with some opportunities for the students to practice the use of this new notation and its connection with types of paper.

Results The children know how to find equivalent pairs. They know how to compare the thickness of sheets (many by two methods). They have a strategy for ordering the pairs, using these comparisons. They know how to use a fraction to designate the thickness of a sheet of paper and how to find equal fractions. They do not know how to check the equality of two fractions in the general case.

They know how to do all these things within a situation. At this particular moment it is not possible to detach a question from the situation and pose it independently. Hence these results cannot yet be built on as knowledge that has been acquired and identified as such by the student.

Module 2: Operations on Rational Numbers as Thicknesses

The next five lessons constitute the second module, which deals with operations in the context of the sheets of paper.

Lesson 1: The Sum of Thicknesses (Summary of Lesson)

By way of motivation for introducing operations, the teacher asks students to consider individually and then discuss with each other the issue of whether the "rational thicknesses" they invented in the previous lessons are numbers. In general the conclusion is that if you have 8/100 the 8 and the 100 are numbers, but 8/100 is two numbers. The teacher points out that we might be able to regard them as numbers if we could do the same things with them that we do with numbers, and asks what those things are. Responses generally include "count objects with them", "put them in order" and "do operations like addition, subtraction, multiplication and so on with them." Quietly tabling the first of these for the moment, the teacher presents the suggestion that to decide whether these are numbers they need to try to do some operations with them.

The first project is to make "cardboard" by sticking together (or rather pretending to do so) a sheet of type A paper (thickness 10/50 mm) and a sheet of type B paper (thickness 40/100 mm.) "How thick do you think the resulting sheet of cardboard will be?" Students agree that that thickness will be 10/50+40/100 mm, and most agree that the result will be 50/150 mm, though a few have some doubts about that. After a short discussion, whatever its outcome, they set out to verify the results. The teacher has them count out 50 A sheets and 100 B sheets and begins gluing (that is, pretending to glue) them in pairs, continuing until students realize that a problem is developing and stop the process. Offered an opportunity to correct their proposed solutions, most go immediately to the correct solution. Most are, in fact, so confident that they declare verification unnecessary, but the teacher does it for the sake of the others, counting out 50 more sheets of type A paper and combining the resulting piles. The stack may measure 59 mm or 61 mm, but this they have already learned to deal with.

They then practice by adding some other pairs and triples of fractions, and observe that they are now capable of adding any fractions they want.

Remarks on This Step: The Choice of Values

To offer at this particular moment the sum of two fractions with like denominators would be a didactical error. Certain teachers have tried it with the hope of obtaining an immediate success for everyone. They wanted to avoid having students have the double difficulty of having to decide to reduce to the same number of sheets and doing it in such a way that the sum of the numerators, that is, the thicknesses, would make sense. Doing so gives the children justifications which are easy to formulate and learn, which facilitates the formal learning of the sum of two fractions (we know how to add two fractions whose denominators are the same, so what is left for us to do in the general case is to reduce it to having the same denominators before performing this addition).

But this method gives inferior results. Only the students capable of comprehending simultaneously and immediately both the general case and the reasons for the apparent ease of the particular cases were able to avoid difficulties in developing a correct concept of the sum of two fractions. They were then able to reason directly or make rapid mental calculations. The rest were distracted by the apparent ease of carrying out the action from the pertinent questions (such as why the denominators can't be added) and the efforts necessary to conceive of and validate the concepts. They were invited to learn a method in two stages, with the possibility of some false justifications for the first stage (if I add three hundredths and five hundredths that makes eight hundredths, just the way three chairs and five chairs make eight chairs.) They first learn that it is possible to add fractions which have the same denominator, and how to do it. They also learn that it is not to be done, or can't be done, if the denominators are different (you can't add cabbages and wolves!) Then they learn to solve the other cases by turning them into the first case, not because of the meaning of this transformation, but because it works. The economy of this process is strictly an illusion, because there is no representation to support the memorization. It will furthermore require a large number of formal exercises to make the process stick and to make it possible to distinguish it from other calculations. Some students never do get it figured out.

Using different denominators, on the other hand, all the children are able to come up with the concept and solidify their representations with experimentation and verification in a way that makes any formal teaching unnecessary.

Delaying the introduction of algorithms can, at times, be of considerable benefit to the development of concepts.

> **Results** All the children know how to find the sum of two or more fractions if they represent paper thicknesses and if the conversion to the same number of sheets is "obvious" (one denominator is a simple multiple of the others). Many would be able to work out a strategy in the case of any two fractions, but no method has been formulated, much less learned.

Lesson 2: Practicing the Sum of Thickness. What Should We Know Now?

The next session comes in two sections which look similar but have quite different functions. Each contains a series of problems. Those in the first section are designed to let the children make use of what they have figured out in Lesson 1, both in order to solidify that knowledge and to extend the range of mathematical activities it can be used for. The first problems are strictly review. The teacher writes up several pairs or trios of fractions to add, walks the class through the first one, speaking in terms of thicknesses of the two papers, and turns them loose on the rest. The next problem is to find the thickness of a sheet obtained by gluing together one of thickness 4/25, one of 18/100 and one of 7/50. Following that, they work on 8/45 + 5/30. The last in this set returns to asking the question in terms of the sheets themselves: "A woodworker is making a collage for a piece of furniture. He glues together three pieces of wood of different thicknesses: 40/50 mm, 5/25 mm and 6/10 mm. List these woods in order of thickness, then say how thick the resulting sheet will be."

In each case, the problem or problems are to be solved individually, then to be presented to the class for discussion and validation. Included in the discussion is the possibility of having several correct routes to the same solution.

The object of this phase is to permit the children to make use of the procedures they discovered in the previous session, to generalize them and make them more efficient. That is, to let them evolve.

This session is thus neither a drill nor an assessment. The teacher does not pass judgment on the value of the methods used, nor at any moment say which solution is correct.

For each exercise, she organizes and facilitates the following process:

Individual effort but for collective benefit
Collection of results
Comparison of methods
Discussion and validation by the students

A method is accepted if it gives a correct solution (thus becoming an "acknowledged" and correct method), rejected if not. Among the methods that have been accepted, remarks on length or facility of execution, which the teacher solicits, do not become judgments of value that the child can confuse with judgments of validity. On the contrary, the teacher sees to it that the child takes part in the debate, has a result to offer, is able to discuss his methods and state his position relative to his own knowledge.

The immediate collective correction and rapid discussion of the problems is thus indispensable. It enables the teacher to know each child's stage of assimilation and what she is having difficulty with. The whole class can take part in each student's effort.

The second phase of the session is a set of individual exercises for drill and assessment. It has a classic didactical form: written questions to be answered individually and turned in for correction (outside of class) by the teacher. The problems represent each of the levels of operation with fractions thus far obtained – ordering of fractions with unlike denominators, addition of fractions with denominators which are like, or one of which is a factor of another, or which require a common multiple.

This frequently results in some rather poor papers, especially since part of its function is to accustom students to the as yet unfamiliar task of producing mathematics for which they have no immediate feedback.

Results This lesson gives lots of opportunity for the exercise of mental calculations with two digit numbers (double, half, triple, multiply by 7). All the children know how to organize and formulate their method for finding the sum of several fractions. They start by trying to reduce them to the same denominator (though the term itself has not been introduced.)

The search for a common multiple has been practiced in many ways (despite the rarity until this moment of occasions for doing so.) Many of the students have begun to work out strategies for a systematic search, such as listing the multiples and comparing the lists, or in the case of small numbers even multiplying the denominators.

Not one of these strategies has been identified as stable, much less learned.

Lesson 3: The Difference of Two Thicknesses (As Measure)

The next session proceeds to the subtraction of two thicknesses. It requires more types of paper, with thickness ranging up to that of heavy card stock, but only one sheet of each of them (for demonstration purposes.)

The lesson starts with a rapid discussion of the problems handed in the day before. Only the ones where errors were made need be mentioned, and the teacher needs to restrain herself firmly from letting the discussion of the common denominator in the last problem result in one of the methods taking on the status of Official Method.

The next stage begins with a swift return to the initial situations: what does 8/50 mean? (The thickness of a sheet of paper such that you have to have a stack of 50 of them to measure 8 mm.) And what does 8/50+6/100 mean? (The thickness of a sheet made by gluing together an 8/50 thick sheet and a 3/50 thick sheet.)

Remark: It is often useful to insert a reminder like that of preceding situations, for two essential reasons:

In the first place to allow children who have some difficulties or are a little slow to be more thoroughly involved in the present lesson;

Furthermore to allow children who have been absent to understand what happened in the previous lessons and be able to participate in the following one.

The teacher then writes on the board

$$8/50 - 6/100$$

and asks the class what that might mean and how to carry out what it says to do.

This launches a discussion that starts with a predictable set of misinterpretations and arrives fairly swiftly at the realization that it is the card stock that is the very thick one, and it is made up of the thin one glued to one of unknown thickness. With a drawing on the board to represent this combination of sheets and the equation $6/100 + \underline{\hspace{1cm}} = 8/50$ beside it, the students are turned loose to work individually on finding just what that unknown thickness might be, and how to verify their results.

The resulting discussion includes many variations, a number of them correct. Students who have not succeeded give the results they got and say whether they are too large or too small.

Next the class interprets and solves $4/15 - 1/15$.

The problem $4/50 - 3/40$ is launched by getting the class to state the need for a common denominator, then left for individual work. Then for a final problem, worked individually, they take on $12/8 - 2/5$.

Results Once again the reduction to the same denominator has given rise to the search for a common denominator. The children still do not know a systematic way of doing it, but are making progress in the sense that more of them recognize more swiftly at what moment it needs doing, and regularly use speedier methods: mental calculation, products of denominators, intuitive searching for the least common denominator.

Lesson 4: The Thickness of a Piece of Cardboard Composed of Many Identical Sheets: Product of a Rational Number and a Whole Number

This lesson requires ten sheets each of four highly distinguishable types of paper, each with a known thickness. Students are set up in groups and each group is assigned a single type of paper.

They are to determine the thickness of a sheet made up by gluing 3 sheets of their own paper together, then 5 sheets, then 20, 100 and 120. Each group figures out the thickness for each stack of their own sheets, then writes the results on the blackboard. Each group then checks one other group's results and either signifies agreement or supplies an alternate answer. Enough students are solidly in control of the material so that the ensuing discussion produces a general agreement, and the chart of values can be successfully corrected.

Examples of the students' debates.

For the first question, group 1 answered	While group 2 wrote
$3/19 \times 3 = 9/57$	$3/19 \times 3 = 9/19$
Their calculation is based on the following schema:	Discussion: Group 2 answers that the thickness of the cardboard can be found by addition
	$3/19 + 3/19 + 3/19 = 9/19$
	They support their result with the following schema which is standard for them at that point:
A student comments that 3/19 and 9/57 are equivalent, so the thickness is the same! It hasn't been multiplied by 3	

The final phase of this session is a comparison of the thickness of the various cardboards with 1 mm. The teacher chooses one of the thicknesses in the chart, for instance 57/35, and asks the students whether they have any idea how thick that card really is. Is it thicker or thinner than 1 mm, or equal to it?

In groups of two or three, students set to work. A lot of them take out their rulers to have a more precise idea of a millimeter. Some work out elaborate approximations, many point out that 35 sheets would make up exactly a millimeter, so 57 of them must be thicker than that ("but not 2 mm thick!") and a few are completely bewildered. After a certain amount of discussion of this particular thickness, the assignment becomes: "Look at the chart and see what else you can say about the thicknesses." This gives rise to a lively discussion and a lot of joy in discovery.

Remark: This last part proceeds informally and spontaneously, for the pleasure of exchanging and discussing ideas without any pressure from the teacher. The teacher listens to the remarks and says nothing unless the students ask him to clarify or explain something.

It is essential to emphasize the fact that the teacher has not set out any contract of learning or acquisition. Some children may take the analysis of the situation a huge distance and make subtle, profound remarks. Others have intuitions which they are unable to communicate. These "discoveries" meander a bit, but it doesn't matter – the jubilation of the ones who have found something wins over the ones who listen, approve, look at them in incomprehension or contradict them. Anyone can advance a notion or even say something that proves to have a major glitch. The teacher restricts himself to making sure people take turns, without interfering with the order or the choice of speakers, in order to maintain the group's pleasure in this game. To do that, he has to register his own pleasure, but make sure that his pleasure is not the children's goal.

He takes note of errors and difficulties without trying to correct them right away. If no one notices them, then in general an explanation at that point would do no good. The teacher has to consider it as an obstacle which needs to be taken up later in a prepared didactical activity.

Frequently after a moment a student notices the error and the debate revives.

Obviously, it has to be clear that the teacher's silence doesn't indicate either acceptance or rejection. And it's not enough to **say** it – he has to **do** it.

Results The children have learned to multiply a fraction by a whole number and to distinguish between this operation and the calculation of an equal fraction. The comprehension of this distinction is essential for what follows. When the children begin to make frequent and varied calculations in more complex problems they will tend on their own to automate their procedures. The initial distinction enables them to do so without losing track of what they are doing and hence to correct the errors that are bound to turn up. Many have begun to envisage the comparison of fractions with natural numbers, a question to which they will soon return. Certainly all of this remains connected to the representation of the fractions by thicknesses of paper.

Lesson 5: Calculation of the Thickness of One Sheet: Division of a Rational Number by a Whole Number

First, the students remind themselves how to multiply by figuring the results of gluing together 5 sheets each 3/9 of a millimeter thick. Then they are presented with:

> "I've glued 9 equally thick sheets of paper together and the resulting card is 18/7 mm thick. What could we ask about it? (the thickness of each sheet.) Can you figure out the thickness? If so, write it in your notebook."

Individual work very swiftly produces the correct result and reasoning.

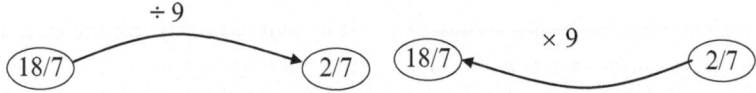

This requires a little delicacy in handling, since they only know for sure that division is defined between whole numbers, but the idea certainly needs confirming, especially after students observe that the operation here can be successfully inverted with a multiplication by 9.

The major point to emphasize is that it is the whole fraction (the thickness) which is to be divided, not just the numerator or denominator. This becomes clearer with the next situation:

"Now I've glued 9 other equally thick sheets together and made a new card. This one is 12/7 mm thick. Can you find the thickness of each of the sheets I glued together?"

Students know how to divide whole numbers. They want to apply the same technique to divide rational numbers by whole numbers. The teacher points out that it might not be the same operation, but accepts it after they compare the properties.

Two out of five groups give the following two-stage response:

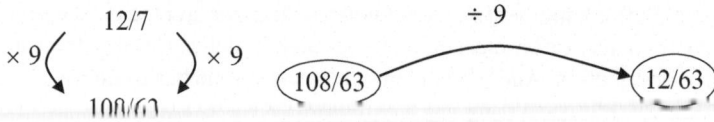

Students work in groups of 2 or 3, then share their results. Since two of the most accessible solutions are multiplying the numerator and denominator by 3 and multiplying them both by 9, the resulting discussion is likely to include a brief furor until somebody observes the equivalence of 12/63 and 4/21.

The final activity is to work individually on (13/5) ÷ 9, first giving it a meaning, then calculating the result. Students tend to bypass the former and work on the latter, which means the teacher has to lean on them to write the sentence in question. After 5 min or so, the teacher stops the work and sends one or more students to the board to write up their solutions. By and large they multiply by 9/9 and then divide the numerator by 9. Only occasionally does somebody observe that the only thing that has happened is that the denominator has been multiplied by 9, and the level of generality of this observation remains undiscovered.

Results Even though most of the children have carried out the operations brought up in this lesson, and have understood the meaning of their work at the moment and in the particular case, there is no guarantee that they will know afterwards how to divide a fraction by a number. But they will find similar situations often enough to develop their methods of calculation, refine them, become confident with them, and hence learn them.

This lesson will enable them to take on these new situations and to understand them without calling forth a reduction to a procedural technique.

Lesson 6: Assessment

The module finishes with a set of problems for a summative evaluation:

1. Put the following thicknesses in order from thinnest to thickest

   35/100 mm; 3/5 mm; 62/97 mm; 5/25 mm

2. Find the sums of the following thicknesses:

   15/100+22/100+62/100
   7/25+14/50+45/100
   3/12+1/4+2/3
   5/8+13/88

3. A piece of cardboard is made by gluing together five identical sheets of paper, each 3/25 mm. thick.

   (a) How thick is the cardboard?
   (b) Is this cardboard thicker or thinner than a millimeter?
   (c) How many sheets would it take to make it thicker than a millimeter?

4. A piece of cardboard is 7/25 mm thick. It is made of eight identical sheets of paper glued together. How thick is a single one of those sheets?

5. Find two fractions equal to 3/18

Note: The fractions are sometimes written with a horizontal fraction bar and sometimes with a slanted one.

Module 3: Measuring Other Quantities: Weight, Volume and Length

The third module (three 1 hour sessions) extends the students' thinking beyond sheets of paper, with the objective of giving them enough similar experiences to make generalization plausible and legitimate. The students use the method of commensuration for three different amounts: volume, mass and length (Fig. 2.2).

Lesson 1: Making Measurements

The first lesson requires a considerable collection of materials:

To measure weight, a balance beam and five different categories of nails;

To measure volume, five small glasses of different sizes, one colored glass to serve as a unit and two (largish) test tubes, one of them with a sticker on it so that they can be distinguished [Note that it is better to do these measurements with fine sand than with water!];

To measure length, strips of construction paper of equal width but different lengths, a single strip of gray cardboard (same width, yet another length) to serve as the unit and a big piece of poster paper to work on.

The glasses, the nails and the strip lengths need to be chosen in such a way that none is an integer multiple or divisor of the unit. For instance, seven nails of one sort might have the same mass (balance on a scale) as eleven of another. If the first serves as a unit, the second weighs 7/11 unit. Similarly, if the content of three "unit" glasses emptied into one tube comes to the same height as the content of five glasses A emptied into the matched tube, then glass A holds 3/5 of a unit. The lengths of the paper strips are between 3 and 30 cm, but they are not any exact whole number of centimeters.

The unit chosen is neither the largest nor the smallest of the available objects. The problem of approximation and precision has to be solved by student agreement with the help of the teacher. An example of how 3/5 of a unit appears for each of these measurements is shown in Fig. 2.3.

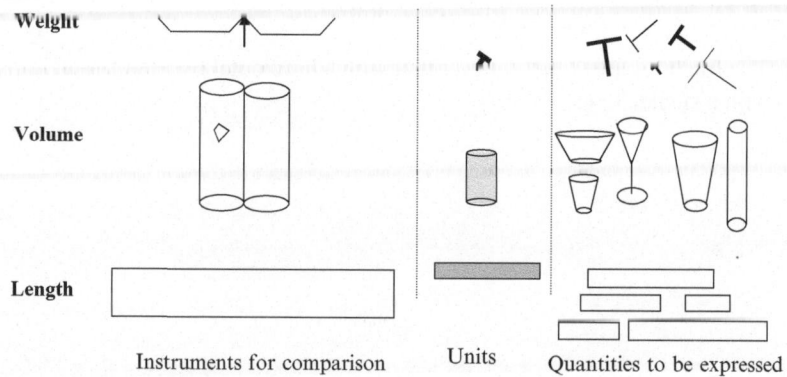

Fig. 2.2 Materials for measuring weight, volume, and length

Fig. 2.3 X is 3/5 of the unit

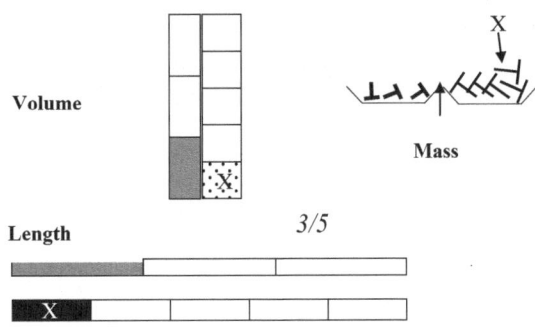

Volume

Mass

Length   *3/5*

The Situation

For a class of 24, the teacher sets up 12 stations – for each of the categories of material, two pairs of stations. The class is divided into six teams, one team for each pair of stations. Each team then splits itself between its two stations, which are at some distance from each other. At their stations, they label each size of object, with the unit having the label U. They then work on figuring out a way to designate the measure of each size of object and on writing a message to indicate the measure of one particular object. When both halves of the team have produced a message, they exchange their messages and try to interpret the message they have received. Then they meet to ascertain the success or failure of the communications and to discuss the best form of communication.

At the end of one such cycle, the team moves on to a different category of object. Each team thus needs to carry out three cycles in order to explore all the types of material. Since each cycle has three parts (inventing a message, interpreting a message and discussing the result) the lesson presents a considerable challenge to the teacher. He must adapt to the students, stimulate them without imposing tedious reproductions, get them to work seriously, with a focus on the task at hand and on the understanding needed to accomplish it. Creating and maintaining the enthusiasm and focus require an exceptional pedagogical performance on the part of the teacher: great rigor to keep the activity rapid and efficient and great flexibility to keep from requiring the completion of tasks that are no longer of interest.

In point of fact, there is no need for every experiment to be carried out by every child. The similarity of the methods swiftly leads the students to re-use commensuration with the glasses and the nails. As the lesson progresses, they get more and more interested in what happens with the length measurements, for which the students soon want to proceed in a different way, but don't know how to write the procedure because it is not a commensuration! The students have no mathematical difficulty with the measurement of mass and volume and do not measure all of the quantities available. The rhythm accelerates. The last cycle is abbreviated. All of the students are set to take an interest in the next day's lesson on measurement of lengths and the comparison of commensuration with subdividing the unit.

Conclusions

The class concludes that their codes for commensuration can be used to measure weight, capacity and length. The session finishes with some practice questions, e.g. "What does it mean that this glass has a capacity of 3/4 of the unit? that this paper strip is 17/25 as long as that one? that this nail weighs 20/75 of a unit?"

Lesson 2: Construction of Fractional Lengths: A New Method Appears

In the previous session, the children attached numbers to sizes (they designated a measurement). In this session, they construct objects whose measurement in terms of a unit is given (i.e., they realize a size). The class deals only with lengths. One reason is that it is difficult to construct volumes and masses of a desired size starting with a random unit. But there is another reason: the teacher wants to get the students to discover another way of defining fractions.

The students have already known for a couple of years how to use the usual method for measuring length in the metric system. In this system the method of measurement always consists of comparing the length to be measured with a whole number of smaller units. To increase precision, one switches to a unit that is ten times smaller. And for practicality, rather than re-measuring the whole length, one measures only the piece that sticks out beyond the part that could be measured with the previous unit, as one does with the remainder in division.

The method we want to induce consists of (for instance) realizing 5/4 by first "partitioning" the unit strip by folding it in quarters, then repeating the resulting quarter-length strip five times.

Materials: Strips of construction paper, all the same width (around 2 cm)
12 unit strips (gray) 20 cm.

Four identical sets of six strips (green) whose lengths are respectively:
5 cm (1/4 unit), 10 cm (1/2 or 2/4 unit), 15 cm (3/4 unit), 30 cm (3/2 or 6/4 unit), 35 cm (7/4 unit), and 45 cm (9/4 unit)

Four identical sets of six strips (blue) whose lengths are respectively
4 cm (1/5 unit), 8 cm (2/5 unit), 16 cm (4/5 unit), 24 cm (6/5 unit), 28 cm (7/5 unit), and 36 cm (9/5 unit)

Four identical sets of six strips (yellow) whose lengths are respectively
2.5 cm (1/8 unit), 5 cm (2/8 unit), 12.5 cm (5/8 unit), 17.5 cm (7/8 unit), 22.5 cm (9/8 unit), 27.5 cm (11/8 unit)

Strips of poster paper 50 cm long and 5 cm wide
Long strips of construction paper, all 2 cm wide
Scissors.

The unit strip should be clearly distinct from the strips to be measured, because since measurement by commensuration consists of laying multiple copies of one strip beside multiple copies of the other, the strips are treated identically in that process. The students naturally tend to confuse 4/5 and 5/4 at first. That is one of the inconveniences of commensuration.

Communication Game and Building Lengths Corresponding to a Pair

Assignment

The class is divided into 12 groups of 2 or 3 children. Each group has 1 unit strip and 1 set of 6 strips of the same color.

"Each group is to find fractions representing the lengths of their six colored strips using the (gray) unit strip and write all of them on the same message pad. So each group starts off as a message-sender.

Each group will receive a message from another group. At that point you all become message-receivers. You are to cut strips of white paper in the six lengths indicated on your message.

Next, each receiver-group will meet with the group that sent the message they decoded and verify together (by superposition) that the white paper strips are indeed identical to the ones used to produce the message. If they are identical, the message-senders are winners."

For convenience, it is the teacher who passes the messages. Groups need to receive messages from other groups whose strips are of a different color (and hence a different set of lengths).

Strips of white paper and scissors are given out at the same time as the other strips.

Development

Initially, students use the method that was inaugurated with thicknesses and generalized to masses and volumes. To realize a length of 5/4 of the unit they lay five units end to end and then try to divide the result in four.

For that they make a guess at an approximate length, repeat it four times and compare it to the length of five units. If the result is too long, they snip off a bit and try again. Some of them observe that the strip they are trying should be shortened by a quarter of the extra length.

They verify that their message was well written and well read and that the construction requested was correct, by superimposing the resulting strip on the original. This method calls for a good mastery of the definition and a certain mental flexibility in applying it, but the students have used that a lot in concrete operations.

When the process of using commensuration results in multiples that take too much space, some of the students think of using the method of dividing the unit. They think of it particularly readily when the natural numbers in the ordered pairs

Initial method: commensuration

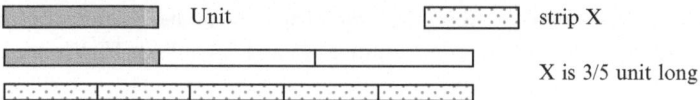

New method: breaking up the unit to produce an intermediate unit that can be used in the familiar way

Fig. 2.4 Examples with 3/5

are simple – 2, 3 and 4 – and the denominator is 2 or 4 (a power of 2): for example the lengths of 3/2 or 3/4 inspire them a bit better than 2/3 or 4/3. They fold the unit strip in two or in four. And they can express those measurements orally by halves or quarters using references to everyday life.

Once they have launched the idea with powers of 2, they progress to other denominators, like 5.

But they can't justify the length directly, with their initial definition. They can only do it by putting five copies of Strip X beside three copies of the unit strip and showing that the lengths are the same, that is demonstrating their equivalence. This will be the subject of the next lesson. Until then, all they can do is write the length of A (1/5) and use multiplication (which they have already encountered): $1/5 \times 3 = 3/5$, trusting to the similarity in writing to carry them through.

Lesson 3: Comparison of Methods, and Demonstration of Equivalence

Summary of the Lesson

This session begins with a follow-up discussion in which by use of the solutions written on the board by the children and a process of observations (by students) and (student-proposed) verifications the teacher guides the class to a conviction that this method of "intermediate units" provides a general solution. For instance, the students can prove step by step that subdividing the unit gives the same result as commensuration because they can write the steps (see Fig. 2.4).

Lesson 4: Fractions of Collections

The follow-up is a pair of problems to be worked on individually and then discussed:

A cloth merchant sells first half of a piece of velvet cloth and then a quarter of the same piece.
What fraction of the piece is left at the end of the day?
The piece was originally 24 m long. What is the length of the remaining piece?

Claude has a bag of marbles. In the course of a game he loses first 2/3 and then 2/9 of his marbles.

What fraction of his marbles has he lost?
What fraction of his marbles does he still have?
At the beginning of the game, he had 63 marbles in his bag. How many does he have at the end of the game?

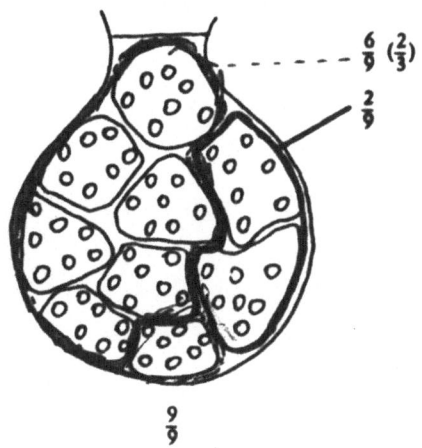

We expected the problem with the marbles to be a good deal more difficult for the children than the one with the cloth, because it combines a variety of difficulties (it requires adding losses; the number is too large for commensuration to be realized; etc.) We observed moreover that in this situation the children didn't recognize the definitions of fractions that they had previously used with continuous quantities. They especially had difficulty conceiving of the bag of marbles as the unit. But manipulation of discrete quantities eventually enabled them to establish a correspondence with long division that they knew.

This lesson completed the study of fractions as measurements.

Results of This Portion of the Sequence of Lessons

The children can fluently use ordered pairs of numbers to express measurements. In fact, they have solved the practical problems of manipulation, measurement, comparison of sizes, evaluation of sums, equalities, multiplication by a whole number, etc.

The culture at large makes use of an imposing vocabulary to conceive of and express fractions. Three heavy-duty difficulties for the teachers are:

Limiting their vocabulary to terms that have been defined and understood
Limiting their explanations to ones made possible by previous lessons
Avoiding analogies and metaphors.

One question was raised every year: Are there commensurations that will never work? That is, are there pairs of objects for which no amount of repetition of each one will ever result in the exact same measurement for the two?

Module 4: Groundwork for Introducing Decimal Numbers

The first three lessons in module 4 are review lessons, so we present only the fourth and last lesson.

Lesson 4: Whole Number Intervals Around a Fraction

First Phase: Introduction to the Game

(a) Instructions and game

"We are going to learn a game that is to be played by two teams. But to understand the rules well, first we will have two students play it at the board in front of the whole class.

Player A will chose a fraction that is somewhere between 0 and 10 (without saying it out loud). She will write it on a piece of paper which she will put in her pocket.

Player B will try to bracket player A's fraction between two consecutive natural numbers. To do that, he will ask questions. For example: 'Is your fraction between 7 and 9?' A can only reply with 'Yes' or 'No'. B will keep asking questions until he has found two consecutive whole numbers that the fraction is between. At that point, A will show her paper and the whole class will compare her fraction with the interval B found.

After that you're all going to play the game, but this time the players will be two teams each made up of half of the class" (and the teacher swiftly creates two teams).

(b) The playing of the game

Now each team chooses a fraction that all the students on that team write in their notebooks. The students choose a representative to play at the blackboard for them. It is the representatives of each team who take turns posing and answering questions. As they do, they can get help from their team by a discussion between rounds. The first team to bracket the other team's fraction in an interval of length 1 wins.

Remarks

1. The intervals chosen by the two representatives should be written on the board. Note that the class has a previously established convention that intervals are closed on the left and open on the right.

The board is divided in half, one for each team.

For example, if Team B has chosen 25/30 and the representative of Team A asks "Is your fraction between 0 and 7?", he writes:

TEAM A
[0,7)

and then after Team B has replied, adds:

TEAM A
[0,7)   yes

If the student then asks "Is your fraction between 5 and 10?", then when the opposing team says "No", he puts a line through the interval:

TEAM A	
[0, 7)	yes
[5, 10)	no

The convention that intervals are closed on the left and open on the right gives rise to the following:

If the fraction chosen is 25/5, then the interval of length 1 containing it is [5, 6), and if this interval is chosen the opposing team answers that it is "trapped", because it is the left hand end-point of the interval.

If the fraction chosen is 30/5 and the interval requested is [5, 6), then opposing team says "no".

If a team brackets the other team's fraction, it scores one point.

If a team traps the other team's fraction, it scores two points.

2. There is an appearance of strategies in the choice of the intervals. This first game between the teams gives them a chance to develop interesting strategies in their choices of intervals. By and large on the first round the representatives tend to ask questions randomly, often overlapping intervals in ways that make their teams lose.

   Example: First question: "Is your fraction between 5 and 9?" Second question: "Is your fraction between 3 and 9?"

   This produces some lively discussions within the team. Often already on the second round they make use of the binary nature of the situation. For example: "Is it between 0 and 5?" If the representative of the other team says "No", they avoid asking "Is it between 5 and 10?", as they often do in the first round. Often after three or four rounds the students manage to locate the fraction with a minimal number of questions.

Remarks

1. If the instructions, which are long, are not well understood, the team game gives the teacher a chance to explain them better, to check that all students know how to write intervals and that they know how to play the game.

2. The team game needs to be restarted a number of times in order for all of the students to understand the rules (there may well be three or four rounds.)

3. The choice of the fraction at the beginning of a round always produces interesting discussions because students often propose a fraction that is not between 0 and 10. Team mates that disagree, if they want to reject the fraction proposed, have to prove to the rest that it is not between 0 and 10,

4. The students swiftly get to the point of avoiding choosing fractions that can be "trapped", because they don't want their opponents to get two points.

Second Phase: Playing Two Against Two

(a) Presentation
   After three or four rounds of the game in large teams, the teacher puts the students in groups of four, so that they play two against two.

(b) Playing the game
   Each pair keeps notes on a piece of paper both of the fraction it has chosen and of the intervals they have asked about for locating their adversary's fraction. The teacher does not intervene except to settle conflicts or supply clarifying information requested.

Third Phase: Collective Synthesis

(a) Presentation
   During the previous phase the teacher has put the following table on the blackboard

Trapped fractions		Bracketed fractions	
Fraction chosen	Interval requested	Fraction chosen	Interval requested

   She interrupts the game played in teams of four after 4–8 min and asks the students:

"Who trapped a fraction?"
"Who bracketed a fraction?"

   She writes up the trapped and bracketed fractions along with the intervals in which they were placed. All the students check these results as they are written up, under the guidance of the teacher.

Results At the end of this session all of the students know how to play the game, and almost all are able to locate fractions in intervals of length one.

Module 5: Construction of the Decimal Numbers

Lesson 1: Bracketing a Rational Number with Rational Numbers: Chopping up an Interval

First Phase

(a) Presentation of the problem and review of the game in the previous session.
"During the last class we learned how we could locate a fraction by figuring out which whole numbers it was between. Do you think it could be useful to know which whole numbers a fraction is between? Why?"
Sample answers:

1. Bracketing lets us say whether the number is large or small.
   Comparison with whole numbers is useful in measurement and in evaluation.
2. Is bracketing useful for comparing two fractions?

   For example, 156/7 and 149/6.
   First method of comparison: give them the same denominator.
   Second method of comparison: bracket them between two whole numbers.
   Which method is shorter?

3. Bracketing also makes it possible to estimate the sum of several fractions. What interval can one give to the sum when one knows the interval for each fraction?

(b) Instructions.
"We are going to play yesterday's game again. Teams A and B will each choose a fraction and designate a representative who will go to the board and ask questions."

(c) Playing.
The game is played exactly as before until the fractions are bracketed or trapped. But while the fractions are still hidden the teacher interrupts the game.

Second Phase: The Search for a Smaller Interval

(a) Presentation and instructions
"You just bracketed the fractions in an interval of length 1 (for example, [3, 4)). Do you think the fraction you were looking for is the only one in that interval? Find some others!"

(b) Development
The teacher lets the students search individually or in pairs for a minute or two. Then he asks them to come write on the board (or writes himself) the fractions they have found that are in the interval.

The children observe that there are many, and that the interval of length 1 that they found doesn't let them give a precise location for the fraction being searched for. They thus understand – some even say it – that they are going to have to find a smaller interval.

Third Phase: Search for Smaller and Smaller Intervals

(a) Instructions

"We are going to add a new rule to the game: to win, you have to bracket the fraction in the smallest interval you can. So you're going to have to try to find smaller intervals and designate them."

(b) Development

Students work in groups of 2 or 3 (there will be 4 or 5 groups per team). Some of the groups have the idea of writing the end-points of the interval as fractions (for example 6/2 and 8/2 if the interval is [3, 4)) But it also happens at times that a lot of students don't think of it and have difficulties. To avoid discouraging them, the teacher may suggest it to them after a few minutes, which revives their interest.

As soon as they have found and designated a smaller interval, they gather again into their two large teams A and B in which each group proposes the interval it has found. The children on the same team then discuss and agree on which among the 4 or 5 intervals proposed they judge to be the smallest.

Then one of the two representatives of the teams comes back to the board and the game continues:

"Is your fraction between 6/2 and 7/2?" (for example)

To answer the question, the students generally request to get back together as a team.

Remarks

1. To answer the question they often call for help from the teacher, because they can't find the answer or can't agree on it. Some of them think of putting all three fractions (their original fraction and the ones being proposed as end-points of the interval) over the same denominator, others give random answers.

   To sustain the pleasure of the game and the desire to continue, the teacher can aid them by giving a few hints (suggesting a common denominator, for instance, if they haven't thought of that.)

2. It is rare for them to be able in the course of this session to propose more than two intervals. Indeed, the big calculations (which they have not yet really mastered) take a lot of time, because they must:

   • Find smaller intervals and designate them
   • Check to see whether their fraction is in these intervals, which requires common denominator computations that are often complicated
   • Finally, to see which team has won, compare the last two intervals designated.

Few of the students are capable, at the end of this first session, of easily reducing the intervals or of saying whether a fraction is inside of a given interval.

Some strategies observed

1. It rarely happens that in the course of the first game all of the children write the limits of the intervals with denominators 10, 100, 1,000,… That's why the calculations are long and difficult. In fact, one time in a first session a group proposed the interval [6/40, 7/40) to bracket a fraction that was between 0 and 1. And since that fraction was 12/37, it is easy to understand why the children ran into difficulties in calculation!

2. In the course of one first session, one of the teams (A) designated their intervals with factions of denominator 64 because they made binary subdivisions: a group of 2 in this team had initially cut the interval in 2, and then in 4 in designating it. When the team got together the other children said "But we could make the intervals even smaller by continuing to cut them in half!", and they tried successively cutting in 8, then 16, then 32 and stopped at 64, convinced that their interval would be smaller than the other team's.

   At the same time, the other team (B) proposed intervals designated by fractions with denominator 1,000. Why? Because a group of two girls on this team had first marked the interval from one to ten (to look like their rulers, they said!) Then, still working like their rulers, they designated intervals in hundredths, and then in thousandths. Their calculations were done very swiftly!

   When the team got together for discussion, the three other groups, who themselves had made subdivisions of 10, 4 and 2, immediately adopted the subdivision into thousandths.

   When the representatives of the two teams went to propose their intervals, the children in team A were able to respond very quickly. On the other hand, the ones in team B, who were asked questions about intervals in sixty-fourths, had to make long, difficult calculations, which made them say to the others at the end "Next time choose something easier. Ask us easy questions like ours!" The teacher stepped in to ask why it was easier to answer the questions Team B asked than those that Team A asked. Everybody understood that for fractions with denominators 10, 100, 1,000 … the calculations were much easier, and of one accord they requested to play again the next day. During the second session the two teams chose subdivisions of 100, 1,000, 10,000 … but that day one team chose the fraction 14/10, (which was swiftly trapped) and the other 83/9!

Results

At the end of this session, the children understand:

That it is possible to locate a fraction in an interval of length less than 1
That in that interval there are many fractions
That that interval can be reduced.

But depending on the choice of intervals or fractions, more or fewer of the children master the calculations and are able easily to find a smaller interval.

Note

If the game as described is too difficult and too long (which happens in some classes), it is simpler to have the teams play one after the other:

Instead of having to pose questions and simultaneously respond to questions posed by the opposing team, one team chooses a fraction (team A, for instance). The other (team B) asks questions that will allow it to find the fraction chosen by A. Team A answers these questions.

Thus one team has only to find intervals, and the other only to answer questions.

In this case, it is necessary to fix the number of questions for each team (3–5, for example) and to compare the last intervals given. Then the game starts over with B choosing a fraction and A proposing intervals.

Lesson 2: Bracketing a Rational Number Between Rational Numbers, Shrinking the Intervals, and Observing Decimal Filters

First Phase: Return to the Game from the Previous Session

(a) Instructions

The instructions from the previous session are used.

(b) Development

The game proceeds in the same way. (If the two fractions chosen during the previous session have not yet been caught, the children want to continue that same game.)

We need to distinguish between the cases where the students have divided the interval into tenths, hundredths, thousandths, etc. and those where they are still using fractions with a random collection of denominators.

Case 1: Decimal intervals. The game develops more rapidly, and is therefore more engaging because the calculations can be made very quickly and are not an impediment to the development. That makes it possible to play several rounds. There are still two cases:

- The fraction chosen is a decimal fraction. In this case it will swiftly be trapped and the children will want to stop the game and start another round.
- The fraction is not a decimal fraction. The children, who are beginning to master the calculations, ask for smaller and smaller intervals (in general they get as far as 1/10,000 without losing impetus.) But at that moment the team that is looking for the fraction begins to wonder a bit and make some remarks (the other team celebrates.)

This is what happened for the fraction 83/9 mentioned in the previous section, which never got trapped in spite of very small intervals being used. The children said "It must be that it doesn't have a denominator with zeros, so it should be 7 or 8 or 9!", and they wanted to stop and see the fraction – which produced a very animated discussion. Some said it couldn't ever be trapped "because 10, 100, 1,000,... aren't multiples of 9." Others held out for the contrary, saying finding shorter and shorter intervals would surely make it all work out.

The problem remained open.

The children's reactions were exactly the same the previous year when one of the fractions chosen was 22/9.

Case 2: Non-decimal subintervals. If neither team has yet thought of producing subintervals in tenths, hundredths, etc. (which happened one time) the game quickly becomes slow and messy. Before the children lose interest (or become understandably disheartened) it is a good idea for the teacher to stop the game and suggest to the whole class that they think a bit about another strategy. For example, she suggests finding questions (for designating intervals) that make the calculations speedy. After a little time for thinking and collective discussion, if nobody has proposed subdividing in tenths and hundredths the teacher might propose another game in which she herself plays against the whole class:

– Either she is the one who chooses a fraction and writes it behind the board and the students propose intervals (which each one writes in his notebook)
– Or the students choose one together, writing it in their notebooks, and the teacher asks the questions.

In the former case, she asks several children, writes the proposed intervals on the board and only responds to those who have chosen decimal intervals. In the second case, she herself proposes the intervals, and uses only those with denominators of 10, 100, etc.

Remarks

The children's pleasure in the game is renewed and they notice very quickly that the teacher is specializing in certain intervals. They generally remark on it with a comment to the effect that "All you have to do is add zeros!" They see that the game is faster, and hence more interesting.

To keep up the children's interest, one can use other variations, such as having them play one against one and then two against two.

It often happens that the fraction proposed is a decimal fraction like 990/100. Children who first bracket it in intervals of tenths find 99/10. The fraction is trapped, but it is not the one that was chosen. So the children say "It's trapped, but it's not the one!" The representative, with the help of his team, then proposes equivalent fractions until he finds the required form.

Results

At the end of this session, all the children understand the necessity of choosing intervals in tenths, hundredths, thousandths. They easily manage

- Either to trap the fraction (if it is a decimal)
- Or to bracket it in very small intervals (of the order of ten thousandths or hundred thousandths).

Finally, they have become conscious that there are some fractions that are easy to trap and others that are not. Some of them even spontaneously list them.

Depending on the difficulty of the fractions and of the intervals chosen by the children, it is almost always necessary to carry the game over into a second session (at the request of the children, in fact.)

Lesson 3: Representation on the Rational Number Line

First game:

(a) Instructions

"Today I'm the one who is going to choose a fraction, and I will write it behind the blackboard. You are to catch the fraction by proposing intervals. I will only say "yes" or "no".

(b) Development

The teacher chooses a fraction (145/100, for example), and writes it in a hidden place. The children work in groups of 2 or 3 and write the first intervals in their notebooks. Once the teacher is sure that all the groups have chosen an interval, he asks them one at a time.

The children ask: "Is it between 0 and 5? between 0 and 3?" and so on until they have found an interval of length 1 (in this case, [1, 2)).

The teacher draws a line on the blackboard, represents the different subdivisions and asks a child to come show where the fraction is found:

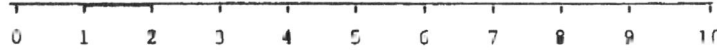

| 0 | 1 | 2 | 3 | 4 | 5 | 6 | 7 | 8 | 9 | 10 |

She draws this interval [1, 2) in color. Then she asks the children to find shorter intervals. At each step, the children indicate the length of the interval (at the request of the teacher).

The game continues until the interval [145/100, 146/100) is proposed, at which point the teacher says "Trapped!"

(c) Many strategies emerge

The students propose intervals in hundredths right off, for example [100/100, 150/100), and then progressively [100/100, 125/100) until they get to [145/100, 146/100)

They start with intervals in tenths. For example:

[10/10, 15/10) bracketed
[10/10, 13/10) the teacher puts a line through it
[13/10, 14/10)
[14/10, 15/10) bracketed.
At that point they propose hundredths.

Each time the children propose a new subdivision, the teacher has them come to the board and write the division points as fractions:

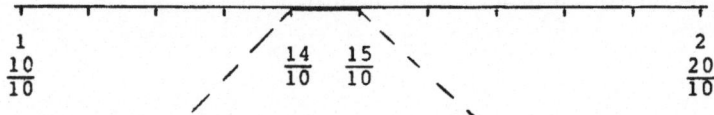

At this stage, the students realize that they are going to have a hard time drawing the division of the interval [14/10, 15/10) into ten equal parts. They propose an enlargement of the interval which they will cut into ten equal pieces. At that point a student will come up and mark both the end points and the intermediate points in hundredths:

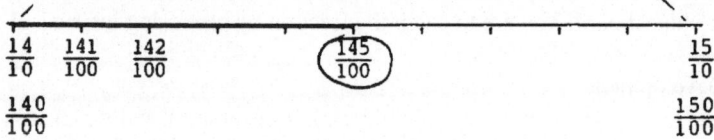

They make new proposals:

[140/100, 142/100);
[142/100, 144/100);
[144/100, 145/100);
[145/100, 146/100) trapped!

Placement on a line

(a) Instructions

"We are going to suppose that this fraction, 145/100, represents the length of a ribbon that we are going to trace in red. So if I put this ribbon along a line marked off from 1 to 10, 145/100 marks the point on the line where the end of the ribbon will be. We are going to put this point on exactly."

(b) Development

This is a collective phase. The activity takes place very quickly in question-and-answer form. On the line drawn on the board, a student comes up and colors the interval [0, 1) in red, then proposes to divide the interval [1, 2) into ten parts, which is also done (either by the teacher or by the student). The endpoints are marked with fractions as they were in the first phase. He extends the red line to 14/10 and then says "We have to cut it in 10 again to have hundredths." The teacher asks what has to be cut in ten. The student shows the interval [14/10, 15/10) and marks the fraction 145/100. He finishes by marking in red the interval [140/100, 145/100).

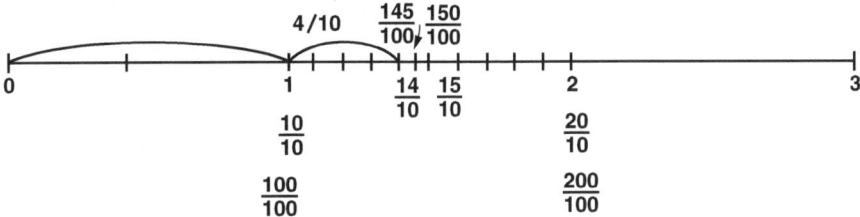

The teacher then asks:

"To measure this ribbon, how many units do we need? How many tenths beyond the 1? How many hundredths?"

And she writes on the board:

Number of units   1
Number of 1/10   4 (so 4/10)
Number of 1/100   5 (so 5/100)

Then she says: "Here is what we measured:"

$$1 + 4/10 + 5/100$$

and asks a student to come to the board and carry out the addition.

The student writes:

$$100/100 + 40/100 + 5/100 = 145/100$$

Remark: The children often say "We took the fraction apart!"

Second game

(a) The teacher proposes that they play another time. For this game the fraction chosen should be a bit different, for example 975/1000.
The game develops exactly as before – the fraction is placed on the line, and then decomposed:

$$9/10 + 7/100 + 5/1000$$

$$900/1000 + 70/1000 + 5/1000 = 975/1000$$

(b) The teacher next asks the children to take apart the fractions that were trapped in the previous activity, for instance 99/10. Everyone writes in their notebook:

$$99/10 = 90/10 + 9/10$$

$$9 + 9/10$$

and puts 99/10 on the line.

Remark: Some of the students notice that the fraction 83/9 that was also chosen in the previous activity can't be placed on this line marked off in 1/10, in 1/100, 1/1000, …

Third game

(a) Instructions

"Do you think now we could guess a fraction very quickly by asking questions about its decomposition? You're going to try to find those questions!"

(b) Development

One student plays against his classmates. He leaves the classroom while the others choose a fraction that they write in their notebooks (243/100, for example).

The child returns and tries to ask his classmates, one at a time, questions that will help him find the fraction very quickly. After a bit of trial and error, he asks (sometimes helped by the teacher) "How many units are there?" "How many tenths?", "How many hundredths?", etc.

His classmates should tell him when he has trapped the fraction. Then he should write it on the board (with the help of the answers he got) and put it on the line.

(c) Remarks

The children note down the information they receive in very different ways. Here are some examples of notations they have used:

Unit rods   2
1/10 rods   4
1/100 rods   3

or

2 units
4 tenths
3 hundredths

or

2
4/10
3/100

The teacher does not intervene in this game. It is the students who protest from their seats if the answers given are not correct or if the student who is looking for the fraction makes a mistake.

The game can be re-played two or three times – the students stay engaged. It's generally the end of the class hour that puts a stop to this game.

The teacher keeps a list of the fractions chosen by the students in this game, because they are going to be needed in the next activity.

Results The students have learned how to put decimal fractions on a number line. Many know how to place them quickly and surely. Some still have difficulties.

They are aware that some fractions can't be put on a line subdivided in powers of ten.

At the end of this activity, they all know how to decompose a decimal fraction and give the number of units, tenths, hundredths, etc.

Lesson 4: From Writing Decimal Rational Numbers as Fractions to Writing Them as Decimal Numerals

Starting a Table

(a) Instructions

The same instructions as before.

(b) Development

A student goes out, her classmates chose a fraction that she is supposed to find by asking the same questions as in the previous activity.

But then the teacher proposes that the information given be marked in the table below (Table A), which will serve for every game.

Table A.

Values of the Intervals	1	1/10	1/100	1/1000	1/10000

For example, if the fraction chosen is 239/1000, the child who is asking the number of units, tenths, hundredths, etc. puts 0 in column 1, 2 in the column 1/10, 3 in the column 1/100, 9 in the column 1/1000 and writes the fraction found in the last column:

Table A.

Values of the intervals	1	1/10	1/100	1/1000	1/10000	
	0	2	3	9		239/1000

One or two more children can play and put their information and the resulting fraction in Table A.

Writing Fractions in Table A

(a) Instructions
   We're going to put the fractions you chose and guessed in the previous session onto Table A.
(b) Development
   The teacher sends several students to the board in turns to write the fractions from the previous game in table A. Then he has them mark other fractions chosen either by the children or by himself (for example 325/100, 1240/10, 85/10000, etc.)

Remark: This phase is collective. All the children participate, either by going to the board, or by making remarks, or by protesting if the one at the board makes a mistake. It should happen quickly like a game.

Other examples are then done individually. The teacher dictates the following fractions which the students put in the copy of Table A they have made in their notebooks:

$$7345/100, \ 7345/10, \ 7345/10,000, \ 7345/100, \ 7345/1,000$$

Passage to Decimal Notation

(a) Information provided
   The teacher writes on the board (away from Table A)

$$7345$$
$$7345$$
$$7345$$
$$7345$$

and asks the class whether they are all the same number. The students reply that written like that, not in Table A, they are all the same number, even though written in Table A they were different numbers. After discussion with the children about the possible means of distinguishing these numbers, the teacher introduces the decimal point.

$$73.45, \ 734.5, \ 0.7345, \ 7.345$$

They immediately note that it is always placed after the units (intervals of length one).
(b) Reading these numbers
   The teacher tells the students how these numbers are read: "73 point 45" or "73 units, 45 hundredths" and has them read several.

(c) Individual exercises for drill and verification

The teacher proposes the following exercises which are done individually and corrected at once. That way she can immediately spot any students who are still having difficulties and help them.

1. Write the following fractions as decimal numbers:

$$245/100, \ 48/1000, \ 2/100, \ 7259/10$$

2. Write as fractions:

$$2.5, \ 145.75, \ 13,525, \ 3.7425, \ 0.1, \ 0.01$$

Results Almost all of the children understand and can write decimal fractions as decimal numbers and vice versa. When the number is written as a decimal number they can say the number of units, tenths, hundredths, etc. This activity gives them hardly any trouble.

Module 6: Operations with Decimal Numbers (Summary)

In the first four of the lessons above, decimal numbers were always written as decimal fractions. In the last one and its immediate sequel, re-writing them in decimal notation becomes the occasion for various exercises in transcription in both directions, and provides the opportunity for them to make most of the common errors arising from transcription and correct themselves using their knowledge of decimal fractions (Lessons 5.4 and 5.5).

Following that, in Module 6 it is time to "redefine" addition for decimal numbers written as what the students call "numbers with a decimal point". After addition and subtraction, multiplication of a measurement number (a concrete number) by a natural number scalar (an abstract number) is easy to understand and carry out, especially making use of techniques of multiplication and division by 10, 100, 1,000. This cycle of six lessons is a welcome one for the students because it takes them back to a domain that they recognize as a familiar one. The many exercises they are given are much easier to carry out and understand, and the classical errors that normally turn up when the operations are carried out mechanically are easy to flush out with the aid of the knowledge they have developed in the previous activities. Students who spent the previous lessons following along on a route being forged by the class that they could not have forged for themselves find that they finally have some material they can handle on their own. It is a joy to discover all at once that the operations are so easy that they can really handle the reasoning to justify them. They credit the relief to the introduction of decimal notation, in which the operations on the rational numbers can be expressed.

Module 7: Brackets and Approximations (Summary)

Division of decimal numbers by a whole number always stops with the units of the quotient. The students only know how to calculate it exactly in the form of a fraction, so that the result is no longer expressed as a decimal number. The next two lessons therefore deal with systematically extending the bracketing of rational numbers between natural numbers and honing the notion of approximation.

At the end of the two lessons, the students try to bracket as tightly as possible the rational number 4319/29. First they extract the whole numbers by a classical division procedure: the fraction is located between 148 and 149 and there remains 27/29. To bracket this number between two successive tenths, they need to know how many 29/290ths (that is, tenths) there are in 270/290 (that is, 27/29). So they divide 270 by 29. This gives them that the fraction 4319/29 is between 148.9 and 149. And they proceed in the same way. They cut the interval [148.9,149) in ten and check how many 29/2,900ths there are in 90/2,900ths, etc.

When they put together on the board in an organized way the sequence of operations that they had scribbled all over their notebooks the children remark that the sequence of successive divisions looks just like a single division that has been extended. There is a small debate before they accept the idea of giving the name "division" to this new operation that enables them not only to bracket a fraction but to determine the "approaching" decimal number resulting from dividing one whole number by another.

The final session is devoted to a mathematical study of the decimal fractions obtained by approximation (i.e., by division) and comparison with fractions. Are all fractions decimals? Do all divisions come to an end? etc. In the course of this lesson, the students carry out multitudinous divisions, but with an eye to studying their properties, not simply as formal exercises "to learn how". The discovery of periodic sequences produces a passionate interest in these instruments for approaching the infinitesimal.

We have now traversed the first seven modules of the Manual. Where have we arrived? On a mathematical front, we are at a point that demonstrates with extreme clarity an aspect of teaching on a constructivist model that opens it to criticism by those who either are unfamiliar with it or in disagreement with it. In terms of institutionalized knowledge — knowledge that could be put on a written test with a reasonable expectation that any student who has been paying adequate attention will be able to answer most or all of the questions – the volume is not particularly impressive. Certainly the students can handle basic rational number operations (all the arithmetic operations with the exception of division by a rational number) very comfortably. A notable strength relative to what one commonly observes is that they are equally adept with proper and improper fractions, On the other hand, in terms of making use of

(continued)

(continued)

rational numbers, their knowledge is still limited to the context of measurement. Similarly for decimal numerals, they can dependably carry out all of the basic arithmetic operations with the exception of division by a decimal numeral, and they can convert back and forth between fractions and decimal numerals, provided the decimal numeral in question is a terminating one. They can also make use of decimal numbers, but again only in the context of measurement. If that really represented the whole of their knowledge, then complaints about the paucity of that knowledge would be entirely legitimate.

What an individual, paper-and-pencil test cannot reveal is the depth of their knowledge and their degree of ownership. Also not susceptible to testing, but nonetheless both impressive and valuable is their level of "community understanding" – the body of knowledge that is accessible when they work as a group, as a result of partial understanding by many students and the capacity of all of them to listen to each other and explore each other's thinking. Thanks to those "invisible" forms of knowledge, they will be enabled in the following modules to expand their individual, institutionalized knowledge dramatically and at considerable speed. They will be able to invent and re-invent the concept of a rational number as a linear mapping until they internalize it, to assemble a collection of observations and partial understandings into some very solid knowledge about division and to use both rational numbers and decimals in most of the standard contexts. To a large extent this knowledge will be institutionalized and testable, though there will, of course, be some speculations and queries left to fuel future exploration.

The remaining two sections of the book will summarize and discuss the modules of the manual covering this last stage in the learning of rational and decimal numbers.

Module 8: Similarity

Lesson 1: Enlargement of a Puzzle

The first situation put to students for study of fractions as linear mappings is the following.

Instructions:

Here are some puzzles (Example: Fig. 2.5 below). You are going to make some similar ones, larger than the ones I am giving you, according to the following rules:

The segment that measures 4 cm on the model must measure 7 cm on your reproduction.

When you have finished, you must be able to take any figure made up from pieces of the original puzzle and make the exact same figure with the corresponding pieces of the new puzzle.

I will give a puzzle to each group of four or five students, and every student must either do at least one piece or else join up with a partner and do at least two.

Development:

After a brief planning phase in each group, the students separate to produce their pieces. The teacher puts (or draws) an enlarged representation of the complete puzzle on the chalkboard.

Strategies and Behaviors Observed

Strategy 1: Almost all the students think that the thing to do is to add 3 cm to every dimension. Even if a few doubt this plan, they rarely succeed in explaining themselves to their partners and never succeed in convincing them at this point. The result, obviously, is that the pieces are not compatible. Discussions, diagnostics – the leaders accuse the others of being careless. They don't blame the plan, they blame

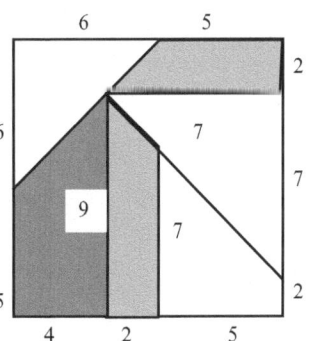

Fig. 2.5 The puzzle

its execution. They attempt verification – some students re-do all the pieces. They need to submit to the evidence, which is not easy to do! The teacher intervenes only to give encouragement and to verify facts, without pushing them in any direction.

Strategy 2: Some of them try a different strategy: they start with the outside square and try adding 3 cm to each of the segments in it. This produces two sides of length 17 cm and two of length 20 cm – not even a square. This is perplexing for the students, who begin to get really skeptical about the plan and often say, "It must be we shouldn't add 3!"

Strategy 3: Another strategy often tried, either spontaneously or after #1 and #2 have failed, is to multiply each measurement by 2 and subtract 1, since they observe that $4 \times 2 - 1 = 7$. This gives a puzzle that is very similar to the original. Only a few pieces don't fit well. So occasionally the students work their way out of the situation by a few snips of the scissors here and there. Even if most of them are aware that they are fudging, a few are convinced that they have found the solution. The teacher, invited along with the other groups of students to confirm success, in this case suggests that the competitors use the model to form a figure with some of the original pieces (such as Fig. 2.6) that cannot be reproduced with the pieces they have produced (Fig. 2.7).

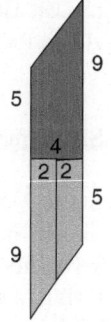

Fig. 2.6 A figure made from pieces of the original puzzle

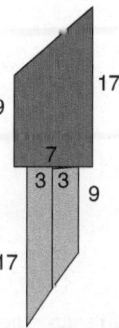

Fig. 2.7 An attempt to produce the same figure after enlarging by ×4-1

To help them see what happened, they can calculate:

$$
\begin{array}{ccccc}
2 & + & 2 & = & 4 \\
\downarrow & & \downarrow & & \downarrow \\
(2\times2\text{-}1) & + & (2\times2\text{-}1) & \neq & 2\times4\ \text{--}1
\end{array}
$$

Remarks: (a) For a variety of social and intellectual reasons, there is a general resistance to the idea of reconsidering the initial procedure. Classes often get quite worked up – lively disputes, accusations, threats – but rarely discouraged.

(b) Occasionally a group succeeds in finding the right process and produces the correct puzzle. The whole class and the teacher take note of the success, but the procedure is examined in the following lesson.

> **Results** All the children have tried out at least one strategy, and most have tried two. By the end of the class, they are all convinced that their plan of action was at fault, and they are all ready to change it so they can make the puzzle work. But not one group is bored or discouraged. At the end of the session they are all eager to find "the right way".

Lesson 2: The Image of a Whole Number

Assignment: "The different procedures you tried out yesterday weren't right, because you couldn't make the corresponding models with the results. You found out that adding 2 or multiplying by 4 and subtracting 1 didn't give the right measurements. Today you are going to try to find the right measurements that will let you make the puzzle right.

Development:

To make things easier, the teacher (or sometimes a student who succeeded with the activity the day before) puts the lengths up as a table:

$$
\begin{array}{l}
4 \longrightarrow 7 \\
5 \longrightarrow \\[4pt]
6 \longrightarrow \\
2 \longrightarrow \\
9 \longrightarrow \\
7 \longrightarrow
\end{array}
$$

Right off the bat somebody always asks for the image of 8 (which is of no use, but which they nonetheless add to the table)

$$
8 \longrightarrow 14
$$

This proposition, which is not rejected, may be what leads to the almost instantaneous appearance of another one: "We need the image of 1!"

"Yes, that would let us find all the others"

The teacher then adds 1 to the table and tells the students to find the measurements. The students work in groups of 2 or 3, all of them having copies of the table in their notebooks. As before, the teacher goes from group to group, encouraging them and answering questions, but does not take part.

Some of the procedures observed:

1.

$$\div 4 \Bigg\langle \begin{array}{c} 4 \longrightarrow 7 = 70/10 \\ \\ 1 \end{array} \Bigg\rangle \div 4$$

$$70/40 \ = \ 35/20 \ = \ 175/100 \ = \ 1.75$$

2.

$$\div 2 \Bigg\langle \begin{array}{c} 4 \longrightarrow 7 \\ \\ 2 \longrightarrow 3.5 \end{array} \Bigg\rangle \div 2$$

Here they are not actually performing a division. They are using cultural knowledge that they have acquired and their explanation is

"Half of 6 is 3
Half of 1 is 1/2 or 0.5
3+0.5=3.5"

From there they continue in the same vein:

$$\div 2 \Bigg\langle \begin{array}{c} 2 \longrightarrow 3.5 = 35/10 \\ \\ 1 \longrightarrow 35/20 = 175/100 = 1.75 \end{array} \Bigg\rangle \div 2$$

3. An alternative for the last step:

$$2 \longrightarrow 3.5, \text{ which they write as } 3.50.$$

To find the image of 1, they write: half of 3 is 1.50, and to that they add half of 50 hundredths, or 25 hundredths. 1.50+.25=1.75;

To find the other measurements, they use either of the following procedures:
Either they multiply the image of 1 successively by 5, 6, 7 and 9
Or they add the image of 1 to the image of 5 to get that of 6, the image of 4 and that of 2 to get that of 6, and so on.

Observation:
One of the children, after having correctly found the image of 1, went on to make all of her calculations using 1.7. When the teacher asked her "Why did you multiply by 1.7 after you had found 1.75?" she replied, "Because I can't measure 1.75 with my ruler because it only goes up to millimeters."

The rest of the class broke in to protest: "Yes, you can! If your pencil is good and sharp you can get very close to halfway between two millimeters." This convinced her, so she didn't do the puzzle with the measurements she had found, and therefore never observed the inaccuracy that would result.

Remark: For many children, measuring 12.25 cm or 15.75 cm gives a lot of trouble that teachers often don't register, but that they ought, in fact, to take into consideration.

Comparison of methods and realization of the puzzles:

As soon as all the groups have found the measurements, they compare and discuss their methods.
The teacher then has them make the pieces and reconstitute the puzzle. (The students would ask to do it themselves in any case.)

Remark: This phase is essential, because for the children it is the only proof that is valid and convincing. But above all, it is source of pleasure and enthusiasm for them: their effort is repaid and they have succeeded.

Results All the children know that the image of a whole number can always be found, and almost all of them know how to find it.

Lesson 3: The Image of a Fraction

First phase: review of the two previous activities:

Assignment: "We enlarged a puzzle. To do that, we had a model on which we knew all of the measurements and we had some information about one of the new measurements: we knew that what corresponded to 4 was 7.

$$4 \longrightarrow 7$$

What did you look for?"

Development: The children briefly recall the activity and the teacher provides a quick overview of all the techniques used. What they needed was the image of 1.

$$\begin{aligned} 4 &\longrightarrow 7 \\ 1 &\longrightarrow ? \end{aligned}$$

She runs swiftly through some other examples: "If 9 goes to 11, what does 1 go to?"
The teacher can send one child to the board or ask them all to work it out on scratch paper.

$$\div 9 \left(\begin{aligned} 9 &\longrightarrow 11 = 11/1 \\ 1 &\longrightarrow 11/9 \end{aligned} \right) \div 9$$

(They often need to review division here, which they do collectively.)

Remark: It is essential for the teacher to pull the class together on a regular basis to remind them where they are: recall or have them recall what problem was posed and what questions that problem gave rise to. They absolutely must know what it is they are trying to solve. The teacher can even occasionally remind them in the course of an activity. The fact is that many children, in the process of working out the intermediate steps of a problem, forget why it is they are carrying out their calculations.

Second phase: Image of a fraction

Assignment: "Now you know how to find the image of any whole number. You also know that you can designate a measurement by a fraction – what did you do that for? (constructing paper strips). Today you are going to try to find the image of a fraction."
    The teacher puts 5/7 in a table of measurements:

$$4 \longrightarrow 11$$
$$5/7 \longrightarrow ?$$

Development:
First she asks the students to think a bit and make sure they all understand the problem posed.
    Spontaneously the children suggest adding 1 into the table of measurements, which the teacher does. She then asks them to find the image of 1.

$$4 \longrightarrow 11$$
$$1 \longrightarrow ?$$

One of the students comes to the board and writes this image of 1, which gives one more rapid review. The new table of measurements now reads

$$4 \longrightarrow 11$$
$$1 \longrightarrow 11/4$$
$$5/7 \longrightarrow ?$$

Remark: The operator 11/4 is not, and should not be, identified. 11/4 is just a measurement.
    At this point, for this particular piece of the problem, the students work in groups of two or three.

Behaviors observed:

1. Many students transform the fraction 11/4 into a decimal numeral: 11/4=275/100=2.75, then stop because they don't know how to multiply 5/7 by 2.75.

$$\times 5/7 \left\{ \begin{array}{l} 1 \longrightarrow 11/4 = 2.75 \\ 5/7 \longrightarrow 2.75 \times 5/7 \end{array} \right\} \times 5/7$$

The majority "trap" 5/7 by doing what they did before[2]

$$\times 5/7 \left(\begin{array}{ccc} 1 & \longrightarrow & 11/4 \\ \\ 5/7 & \longrightarrow & \end{array} \right) \times 5/7$$

But there again they bump into calculations that they don't know how to carry out: the multiplication of two fractions: $11/4 \times 5/7$.

We should point out that a certain number of them do actually write out the correct result: $11/4 \times 5/7 = 55/28$ purely by intuition. Obviously, the result can't be accepted, because they can't justify it at all.

Another frequent event is that they write 4, 11 and 1 as fractions:

$$4 = 4/1 \quad 11 = 11/1 \quad 1 = 1/1$$

and are stuck there.

The teacher goes from group to group, asking questions, giving encouragement. This activity is difficult for children of their age, and they need to be helped along with questions like "What would you do to trap 5/7?" and "What do you think might be another in-between number?"

Fruitless efforts. They remain stuck. So she organizes a collective discussion.

Third phase: the search for an in-between number

Assignment: The teacher first asks the students to look closely at the table of measurements from before:

$$4 \longrightarrow 11$$

$$1 \longrightarrow 11/4$$

$$5/7 \longrightarrow ?$$

She poses the question: "What would make it easy to do the trapping? Think about the calculations you would have to do."

Development: A phase of collective reflection starts up first. The children think silently, then propose things out loud. The proposals are immediately put to the test while the whole class watches.

A certain number of them lead nowhere: a proposal to put 1 in the table in the form 7/7 or 1/1 which makes no progress on the problem because they don't know how to get from 7/7 to 2/7, or from 1/1 to 5/7.

Others, for instance 5 or 1/7, do lead to a possible answer to the question the teacher asked.

---

[2] Some expressions used spontaneously by the children to make themselves understood – either by the teacher or by the other children – are adopted by the whole class and accepted by the teacher. Some terms are thus used that are not necessarily "mathematical" but only serve temporarily for dealing with particular situations. They are not institutionalized, and are therefore later forgotten. Examples: "trap" and "in-between number"

Examples of attempts to check the last two, made at the blackboard by a student or the teacher:

First attempt, using 5:
First 5 is converted to a fraction: 5/1.

Remark: At this point it is once again generally useful to review division by having them quickly carry out small calculations like $25/3 \div 9 = 25/27$; $13/9 \div 5 = 13/45$; $81/13 \div 9 = 9/13$ etc.

(A few students know and recall that they can multiply the denominator by 7 to make the fraction 7 times smaller.)

Second attempt, using 1/7:

$$1/7 \xrightarrow{\times 5} 5/7$$

The children return to their groups of two or three and get back to work on the solution they were working on in the second phase.

Strategies observed:
All the groups make one or the other of the following tables:

Either

$$4 \longrightarrow 11$$
$$1 \longrightarrow 11/4$$
$$5 \longrightarrow$$
$$5/7 \longrightarrow$$

or
$$4 \longrightarrow 11$$
$$1 \longrightarrow 11/4$$
$$1/7 \longrightarrow x$$
$$5/7 \longrightarrow y$$

depending on which of the two proposals they adopt.

Remark: Not all the groups get to the end of the calculations, because the children make mistakes. They have forgotten the techniques they developed during the lessons about operations on fractions. This is perfectly normal, and the teacher should neither worry nor blame the children. On the contrary, this is exactly the moment to re-use the processes they discovered quite a while ago, put them to work and let the children see what the processes are good for.

Collective Synthesis of Methods

First the teacher asks them which groups didn't succeed. She asks them to try to say what messed them up and why they didn't get any result. The children mostly know very well what happened to them: they made mistakes in the multiplication or division of a fraction by a whole number – that's the principal cause of errors.

Remark: Because this is a regular proceeding, the children are perfectly comfortable discussing their mistakes. This is beneficial to the whole class, since exploration of errors can often contribute just as much to understanding as observation of correct procedures.

After that discussion, the teacher sends some students to the board to describe the methods they used:

$$
\begin{array}{lllll}
1) & 4 & \longrightarrow 11 & \qquad 2) & 4 & \longrightarrow 11 \\
 & 1 & \longrightarrow 11/4 & & 1 & \longrightarrow 11/4 \\
 & 5/1 & \longrightarrow 55/4 & & 1/7 & \longrightarrow 11/28 \\
 & 5/7 & \longrightarrow 55/28 & & 5/7 & \longrightarrow 55/28
\end{array}
$$

Exercises for Practice

The teacher adds two more fractions to the table of measurements and the students calculate their image individually

$$
\begin{array}{ll}
4 & \longrightarrow 11 \\
1 & \longrightarrow 11/4 \\
 & \\
7/9 & \longrightarrow ? \\
6/7 & \longrightarrow ?
\end{array}
$$

They discuss their solutions rather than turning them in.

Final step: The teacher inquires: "Does every fraction have an image?"
    After some reflection, the children conclude that you can always find the image of a fraction because you can always multiply and divide a fraction by a whole number.

Results At the end of this session, the children understand that you can find the image of any fraction at all provided you know the image of one whole number.
    They have also all grasped that you have to figure out the image of 1 and of some "in-between number".
    On the other hand, they haven't all mastered the sequence of calculations, and can't all get to the result.

Remark: We emphasize here that this is normal and the teacher shouldn't worry. It would be a serious error to stop and drill the students, because the up-coming activities let them re-use these notions and progressively master them (each child at his own rate.)

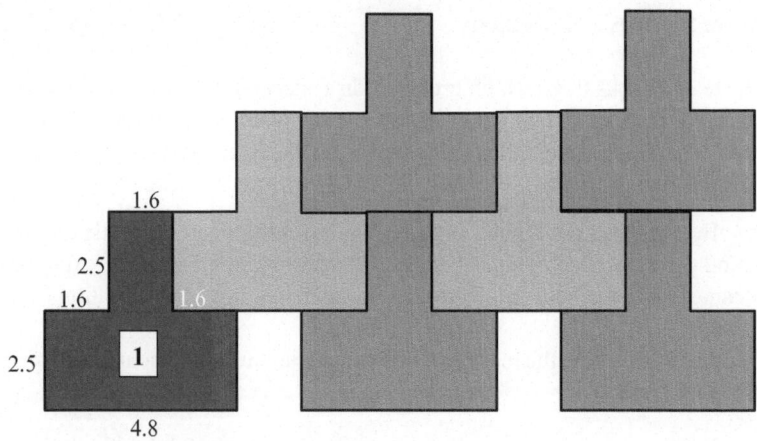

Fig. 2.8 Tesselation to enlarge

Lesson 4: The Image of a Decimal Number

Problem situation: construction of a tessellation

Materials: 3 or 4 cardboard pieces similar to the figure piece marked (1) in Fig. 2.8 or the same figure drawn on the board before the lesson.

Assignment: We are going to make a decorative panel for our classroom. It will be made up of pieces like the one you have, put together like this[3]:

To do that, each of you will make one piece by enlarging the model so that 1 cm on the model corresponds to 3.5 cm on the piece you make.

$$1 \text{ cm} \longrightarrow 3.5 \text{ cm}$$

Development: The children work in groups of three. They start off looking for ways to find the measurements for the piece

$$1 \longrightarrow 3.5$$
$$2.5 \longrightarrow$$
$$1.6 \longrightarrow$$
$$4.8 \longrightarrow$$

Strategies observed:

1. The most common strategy is to convert the decimal numbers to fractions

$$1 \longrightarrow 35/10$$
$$25/10 \longrightarrow$$
$$16/10 \longrightarrow$$
$$48/10 \longrightarrow$$

---

[3]The drawing should be prepared before class, either on the board or on paper.

Then by referring to the previous activity they calculate the images and add them to the table, first calculating that the image of 1/10 is 35/100 and then multiplying by 25, 16 and 48 respectively.

2. Another common strategy is to do the calculations by taking apart the decimal numbers as follows:

For the image of 2.5:

First find the image of 2 by doubling the 3.5.
Next find half of 3.5 by finding half of 3 and then half of 0.5
Then add up all three results.
For the image of 1.6, add the images of 1, 0.5 and 0.1
For the image of 4.8, add the images of 4 and 0.5, plus 3 times the image of 0.1

Remark: This last method is generally used by the children who are very good at mental calculations. Many of the calculations described above are invisible – only the results appear. At the request of the teacher (who goes from group to group and keeps on saying "But how did you get that?") the children consent – often with bad grace – to write them (at times in a highly disorderly way.)

Phase 2: Comparison of Methods

The groups that have found the numbers take turns at the board explaining their method. This gives the ones whose numbers didn't work out a chance to find out what went wrong. When children don't succeed with the activity it is always because of errors in calculation.

Conclusion: As at the end of the previous activity, the teacher asks, "Does every decimal number have an image?" Needless to say, the children respond in the affirmative.

Phase 3: Making the Pieces

Each child makes one or more pieces out of colored paper. As happens with the puzzle activity they are again faced with measurements: 5.6 cm, 8.75 cm, 16.8 cm They also have to use a T-square to make their lines, which adds a second interest to this session: construction of geometric figures.

Results This activity gives the children a chance to re-use procedures worked out in the previous session. It enables many of them to master some difficult calculations that they have previously been unable to carry all the way out.

Lesson 5: Division of a Decimal Number by 10, 100, 1,000, ... *(Summary)*

Still using the set-up of $1 \rightarrow 3.5$, the teacher turns the class loose on finding the images of 1/10, 1/100 and 1/1,000. This presents no difficulties, and very soon the teacher is able to put a collectively produced table on the board. She then writes up the problems and results, including some intermediate problems that the students have produced:

$3.5 \div 10 = 0.35$
$0.35 \div 10 = 0.035$
$3.5 \div 100 = 0.035$
$0.035 \div 10 = 0.0035$
$0.35 \div 100 = 0.0035$
$3.5 \div 1,000 = 0.0035$

The students contemplate this and make observations, checking against different entries: "It's just the reverse of multiplication"; "You have to move the decimal point backwards", etc.

The teacher then leads them to formulate the rule: When you divide by 10 or 100 or ..., you have to move the decimal point as many places to the left as there are zeros in the number.

By way of solidifying the rule and pushing the students a little farther, the teacher gives them some exercises, which are done individually and immediately corrected:

1)	$45.87 \div 1000 =$	2)	$135.9 \times$ ⬚	$= 1359$
	$139.2 \div 10 =$		$4457 \times$ ⬚	$= 485.7$
	$4750 \div 100 =$		$0.129 \times$ ⬚	$= 129$
	$25785 \div 10000 =$		$130 \times$ ⬚	$= 13000$
	$0.08 \div 100 =$		$1675 \times$ ⬚	$= 16.75$
	$0.08 \div 1000 =$		$5.45 \times$ ⬚	$= 5450$

Result As was the case for multiplication of a decimal number by 10, 100, etc., even though the children understand, they make a lot of mistakes with the placement of the decimal point. As a result, it is absolutely necessary to keep on regularly giving them exercises (corrected immediately) so that they master these calculations swiftly, because they are going to need them for lots of other activities.

Module 9: Linear Mappings

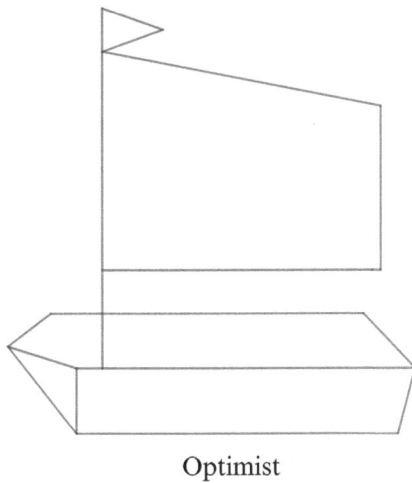

Optimist

All of the sections of Module 9 revolve around reproductions of the drawing above. It was chosen because not long before the time the lessons were given, the class had had a whole month in which after a morning in school, they spent every afternoon together at a sailing school near-by on a boat called the Optimist. This annual event was a source of great pleasure and of great class bonding as well. The basic drawing is on card stock and has the dimensions listed below. In addition there are 11 reproductions also on card stock, with specified ratios of enlargement or reduction.

Lesson 1: Another Representation of the Optimist (Lesson Summarized)

After introducing the drawing and having the children help her label the parts of it, the teacher puts on the board the list of dimensions:

Height of mast	17.7 cm	Length of boom	14 cm
Height of pennant	1.7 cm	Height of hull	3.4 cm
Side of pennant	4 cm	Length of stem	5.2 cm

She then puts up, beside the original, the reproduction that has a ratio of 1.5 to the original. The children observe and make comments: "That one's bigger", "It's not twice as big – it's less than that". They often ask: "Are all the measurements the same?" by which they mean, "Were all the measurements enlarged the same way?" Sometimes they say: "Is it proportional?"

The teacher tells them that they can find that out themselves if they find the measurements of the reproduction. Then she asks: "Would you know how to find the measurements if it were proportional? What information would you need to do it?" The children generally tell her that one is enough.

She sets them up in groups and announces that each group must request in writing the one measurement that it wants. They are then to use that to predict what all the other measurements will be if the enlargement is proportional. Once the calculations are finished, they are to take their rulers up to the reproduction and check the measurements. If all of the actual measurements correspond to their calculated ones they will have the answer to their question.

The groups work together first to find the procedure that will give them the measurements. Then they divide up the measurements – one does the mast, another the boom, etc. As soon as they are done they check their results by measuring. Standard comments: "Yes, it is proportional" or "We blew it! Our measurements aren't the same", in which case they go back to their places and start over. (In one sad case a group that had asked for the measurement of the mast proceeded to add 8.55 to all the measurements, because $26.55 - 17.5 = 8.55$.)

When they are done, they have a collective discussion. First the ones who have had trouble describe where the trouble arose, then the groups that succeeded come to the board and present their methods (one presentation per method).

Some of the methods presented by the children:

First strategy – measurement requested was the length of the boom

The students noticed that $14+7$ (half of 14)$=21$. To find each measurement of the reproduction, they added half of the measurement on the original to the measurement itself	14	21
$3.4 = 3 + .04$	3.4	$3.4 + 1.5 + 0.2 = 5.1$
$5.2 = 5 + 0.2$	5.2	$5.2 + 2.5 + 0.1 = 7.8$
$17.7 = 17 + 0.7$	17.7	$17.7 + 8.5 + 0.35 = 26.55$

Second strategy – again starting with the boom

$$\div 14 \left(\begin{array}{ccc} 14 & \longrightarrow & 21 \\ 1 & \longrightarrow & 1.5 \end{array} \right) \div 14$$

So the image of 1 is 1.5.
For the side of the pennant, then:

$$\times 4 \left(\begin{array}{ccc} 1 & \longrightarrow & 1.5 \\ 4 & \longrightarrow & 6 \end{array} \right) \times 4$$

For the height of the pennant, they write 1.7 as 17/10, then work with the image of 1/10

$$\begin{array}{l} \div 10 \left(\begin{array}{ccc} 1 & \longrightarrow & 1.5 \\ 1/10 & \longrightarrow & 0.15 \end{array} \right) \div 10 \\ \times 17 \left(\begin{array}{ccc} & & \\ 17/10 & \longrightarrow & 2.55 \end{array} \right) \times 17 \end{array}$$

The same strategy was used by some groups who started with the measurement of the side of the pennant. One such group began by calculating the image of all the integers: 17, 3, 4, 5 and 17 and the image of 0.7, 0.4 and 0.2 and adding appropriately.

A third and fourth strategy were developed by groups who started with the height of the mast. One was to multiply both sides first by 10, so as to have whole numbers to deal with. Another was to convert 17.7 to 177/10 and then divide both sides by 177. Both strategies then match those of the second strategy.

Commentary Like any other lesson that involves making actual measurements and comparing them with the results of computations, this one brings up issues related to approximation. The teacher needs to establish very gradually over the course of all such lessons an understanding within the class of how to treat values arrived at by measuring and those arrived at by calculation. Questions of how large a discrepancy is acceptable should be treated case by case, with student opinion always underlying the decision so that they never think the answer is handed down from on high. Eventually error intervals and the algebra thereof should work their way in, but not as a topic in themselves, always as a means of dealing with a particular situation.

Lesson 2: (Summary of Lesson)

The next lesson is highly similar to the first. The only difference is that the new reproduction is the one with proportionality factor 1.4. As soon as the students see it they notice that the first procedure above won't work. They settle down and swiftly work out the new lengths using one or another of the other procedures. After the solutions have been duly discussed, they discuss which one they found the most effective and institutionalize it as the one to be used in the following activities (generally the one that starts by turning everything into a whole number.)

Lesson 3: Lots of Representations of the Optimist (Summary of Lesson)

This is followed by a very challenging lesson that uses a bunch of the reproductions and poses a new problem.

The teacher holds up five of the reproductions, some larger than the original, some smaller, and some very close in size. First she has the class put them in order by size and labels them A, B, C, D, E, and M for the model. She sets up a table with the letters across the top, starting with M, and the six elements of the boat whose measurements they have been working with down the side. She fills in the column of measurements for the model, then the row of stem measurements for all of the reproductions (which settles whether they got the order right.)

Then she says, "I will tell you that one of these has mast length 13.275. Can you figure out which reproduction I'm talking about?"

This is a real challenge to the students, because it is not obvious to them how to attack it – how to identify the relevant variables. For instance, one tactic would be to calculate the ratio of 13.275 to 17.7 (mast length of the model), then calculate the ratio of each of the stem lengths to 5.2 (stem length of the model) and see which one matched. Another would be to calculate the lengths of all of the masts and see which one comes out to be 13.275. Or then again, one could calculate the stem length of the unknown boat and compare it with the given lengths.

Remark Some of the children simply can't get their hands on the problem. This is the kind of situation in which the teacher must firmly resist temptation. If she reduces the scope of the problem by pointing out which numbers are relevant and what to calculate, she will take all the interest out of it. The object of the lesson isn't to accomplish a task, but to determine what it is.

Left to their own devices, the children make remarks like: "The image is smaller than the model, because the mast is 13.275 instead of 17.7" "But not much smaller…" "It can't be E, because E is bigger." This strategy of reducing the field of possibilities provides the opportunity for some good work with ratios: "It's not A, because A is much smaller – the stem is less than half as long as the one on the model."

This way they whittle the possibilities down to two or three. Then "to be sure", they decide to do some calculations. But which ones? What's going to tell them which boat that mast belongs to?

They think it over a while, and after some hesitations and tentative efforts, one of them comes to the board and writes:

Model	Reproduction
17.7 $\longrightarrow$	13.275
5.2 $\longrightarrow$	[ ]

The children work it out in groups of two, and the teacher chooses one to write the correct process on the board. The standard mode of calculation produces a stem length that matches that of boat C.

As a follow-up, the class does a series of problems individually, so that each child can figure out whether he actually understands, and whether he knows how to find whatever measurement he needs.

The rest of the lesson is a swift activity aimed chiefly at the motivation and introduction of a new notation. The original drawing and the six reproductions labeled A through F are still posted on the board. [Note: this lesson write-up is based on a report from a different year from the previous one. In the interim a spare reproduction of the Optimist seems to have turned up!] The teacher brings out four more. One at a time she gives a single measurement from each of them, and the class quickly tells her where to put it. In a short, almost playful time, all 11 are lined up on the board.

Question: "What are we going to call these new ones? We need to be able to talk about them." Many of them suggest A, A_1, A_2, B, etc. The teacher says she has another one that goes between A_1 and A_2, after which the class realizes that letters are not going to suffice. They set to work finding an alternative method. Often one of them will suggest using the image of one, since that way they can tell whether it is enlarged by a little bit or a lot. If none of them thinks of that possibility, the teacher suggests it, and asks them to verify that it provides the information needed. It should (a) let them find the image of any of the measurements and (b) let them put any enlargement she gives them in the right place.

She then has students go to the board to show where to put something enlarged by 1.35, by 1.87, by 0.72 (i.e., shrunk), by 0.29, etc. Then she has them reverse the process and find enlargements to go between ones that are already up there. This they do on their own, on scratch paper. Meanwhile the teacher writes under each reproduction the corresponding image of 1:

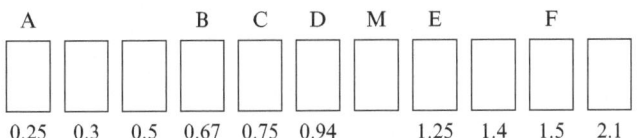

0.25   0.3   0.5   0.67   0.75   0.94   1.25   1.4   1.5   2.1

Enlargements, reductions, 0 or 1?
Next comes a rapid class exchange, launched by the teacher:

"What do you notice about the numbers labeling the reproductions?"
"They're in order from smallest to largest"
"The bigger the enlargement, the bigger the number"
"One of them doesn't have a number – it's the original model."

They decide it ought to have a number, and the class splits between those who propose 0 and those who propose 1.

"All the ones that get smaller have numbers less than 1"
"The numbers bigger than 1 all give enlargements"
"M is in between, so it ought to have a 1."
The teacher steps in with: "If I make a reproduction using 1, what will I get?"
"A reproduction "equal" to the model. It doesn't get smaller and it doesn't get larger."
"That's just it! It does nothing. Enlarging by nothing should mean zero!"

Teacher: "With our convention we have to put 1, but what would a 0 reproduction give?"

$$1 \longrightarrow 0$$
$$2 \longrightarrow 0 \quad \text{etc.}$$

"Nothing!" "A point!" …

With that settled, the teacher goes on to another point: "Do you know how to tell whether the reproduction 0.84 is an enlargement? And 1.10? and 0.01? What would you mean by an enlargement by 2?"

The students answer: $1 \longrightarrow 2$

"How about a reduction by 2? Or an enlargement by 1/2? Is that a contradiction?"

Conclusion: It needs to be called a reproduction $1 \longrightarrow 2$, because the number is all it takes to tell us whether it enlarges or shrinks.

Remarks This lesson is in "Socratic" form – questions and answers. Rather than setting up a Situation of communication, like the ones with which rational numbers as measurements were introduced in Module 1 the teacher contents herself with talking about communication, because here the issue is familiar to the students and nothing new would come of such a Situation.

Students' internalization of the types of Situation that justify the means proposed for managing knowledge is part of the epistemological construction that the teacher is responsible for. This internalization saves time later without losing any of the meaning of the knowledge being created.

Lesson 4: Good Representations, Not So Good Representations

This one requires some special preparation. The teacher needs to make a special transformation of the model that enlarges the model with a horizontal ratio of 1.2 and a vertical ratio of 1.5. The resulting reproduction will be called Z.

As usual, the lesson starts with a review of the preceding lessons, including in this case listing on the board the images of 1 they found for all 11 of the reproductions. This leads up to having the children articulate what constitutes a reproduction that is an enlargement or reduction:

"You take a model.
You measure its dimensions.
You enlarge or shrink all of its dimensions the same way
You get a bigger or smaller picture."

The teacher then divides the class into four groups, with four different tasks. Three of them start with the original model and apply the following three mappings:

$$1. \quad 1 \xrightarrow{\times 2.2} 2.2$$

$$2. \quad 1 \xrightarrow{+5} 6$$

$$3. \quad 1 \xrightarrow{\times 2\, +\, 3} 5$$

Question: "Do these mappings give you enlargements?"

The fourth group gets the new reproduction, Z. Their question is then "Is this an enlargement of the model?"

After the first question is asked, murmurs are audible: "Yes, they're enlargements." There are nevertheless a few who bring up the enlargement of the puzzle: "When we added 3 it didn't work!" From then on the term "enlargement" has some ambiguity for the children, but not all of them can quite say why.

Development: The children in the first three groups decide to calculate all the dimensions using the proposed directions and make the corresponding design with the resulting dimensions. They decide who is to calculate which dimension, then settle down and do it.

Observation: In the course of this phase of the lesson, the children in the groups that were given mappings # 2 and 3 become rather noisy (everybody accuses everybody else of calculating badly): "It's impossible! Our design doesn't look a bit like the model!" "It's not a boat! It's a jam jar!" "The pieces just don't fit!"

They want to quit, and usually call the teacher, who de-dramatizes the situation by smiling and telling them it's OK not to go on with the design.

Class discussion: When they are all done, the teacher has the first three groups put all of their numbers on the board, and the class checks them. Looking at the designs, the children are completely satisfied that #1 gives a proper reproduction. Some of them say "It looks proportional to the model." On the other hand, the designs that correspond to the other two create a lot of laughter. The line segments intersect at weird places, or don't intersect at all. It's impossible to reproduce the shape of the Optimist and use the numbers resulting from the operations in question.

The fourth group has had some problems: their reproduction looks a lot like the Optimist, but they couldn't find a consistent image for 1. Using the mast, they computed the image of 1 to be 1.5, but one of them noticed that the reproduction looked a bit elongated, so they checked the boom. Using 1.5 as the image of 1, the calculated the image that the boom on Z ought to have, and got 21 cm. Then they measured it and got 16.8 cm. They went back and checked all their calculations, but they were all right.

The teacher suggested that they calculate the image of 1 using the boom, so they did that and found 1.2 – not the number they got before! The child who had noticed that the design seemed elongated suggested that they should find the image of the hull with the new image for 1, so they did and it checked out with their measurements. When they presented this to the rest of the class, someone wondered whether all the vertical enlargements might be the same. So the teacher immediately encouraged them to calculate the image of 1 first using the height of the hull and then the height of the pennant. The students noted, "It ought to be the same as for the mast!"

They do a batch of computations and everything works as predicted. Now they know for sure: there is more than one image for 1.

To finish up, the teacher gives them a swift introduction to linearity. She points out that the height of the boat is the height of the mast plus the height of the hull, and writes all three of those measurements on the board. Then she has them compute the images of all three under each of the first three mappings. The first one duly gives images that add up properly, but adding 5, or multiplying by 2 and then adding 3 both give non-matching images. The students remark that it's just like the puzzle situation – you don't get a good reproduction unless you just multiply. Addition just messes things up!

Presentation of information: The teacher confirms their conclusion by telling them that if the sum of the images is equal to the image of the sum, then we say that **the mapping is linear**, or that **the numbers are proportional.**

Game: Invent some reproductions that aren't proportional. Some examples might be:

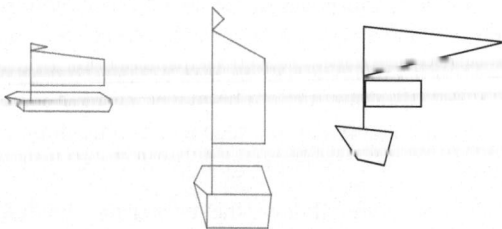

Lesson 5: Change of Model

The whole flock of reproductions is still on the board. The teacher reminds the class of all they have learned working with the puzzle and the boat reproductions.

Assignment: "Do you think we could start with a different model? For instance, suppose I chose C as my model rather than M. Would I be able to designate the enlargement that gives F by the same number? Could you find a number that designates this enlargement?"

Development: The children and the teacher work together – the teacher poses questions and the children answer them.

"We saw that we could get F from M by the enlargement $1 \longrightarrow 1.5$. If I take C as the model, will $1 \longrightarrow 1.5$ still work?"

To make the comparison, the students propose to calculate the respective lengths of the same piece of the design. They choose the boom measurement, which is easy to work with. Some (re-) calculate the image of 1 in F starting from M, thus confirming that it is 1.5. Others calculate the image of 1 in F starting from C, and find that the mapping that takes C to F is $1 \longrightarrow 2$

Students often comment that "It's reasonable for the enlargement to be bigger, because C is smaller than M, so you have to enlarge it more to get F!"

Conclusion

If you change the model, for each figure you can still figure out the image of 1, but it's different than before.

You can't represent a figure by a number unless you indicate the corresponding model. It can be represented by as many different numbers as there are models to choose from.

Reproduction: the action and the image.

"Since we can have different models for the same figure, we can't just put the number that designates the enlargement or reduction under the reproduction. It has to be put between the model and the reproduction in order to designate the mapping." The resulting example is the familiar list of 11 figures, with a curved arrow going from M to each of the others and the image of 1 on the arrow. They are the numbers of the reproduction-action, not of the reproduction-image.

Calculations with Other Images

Assignment: "If we take as a model the reproduction we used to call 0.5 and reproduce it so as to make $1 \longrightarrow 3$, which figure will we get?"

Since the boom on the reproduction in question measured 7 cm, the new one will measure 21 cm, which is the boom of reproduction F.

Next assignment: "What is the mapping that takes us from A, used as the model, to C?"

The teacher gives the boom lengths for both figures and the children compute as they did before. It turns out to be the same as the mapping that took us from the former 0.5 to F.

Another assignment: "If I use the enlargement 1⟶5, I get F. Which one was the model? What table could we set up to find it?"

The teacher and the children exchange propositions, and then the children calculate individually. They use the ratios in F to find the measurements in the model:

$$
\begin{array}{ccc}
\text{Model} & & \text{F} \\
\end{array}
$$

$$
\div 5 \Big(\begin{array}{ccc} 1 & \longrightarrow & 5 \\ & & \\ 1/5 & \longrightarrow & 1 \end{array} \Big) \div 5
$$

$$
\times 21 \Big(\begin{array}{ccc} 1/5 & \longrightarrow & 1 \\ & & \\ 21/5 & \longrightarrow & 21 \end{array} \Big) \times 21
$$

On the model, the boom length is thus $21/5 = 4.2$. That's the reproduction formerly known as B.

Images and Reproductions

To determine a proportional representation, how many reproductions do you have to show? Two, the model and its image. For sure, the same proportional reproduction can make each model correspond to a different image. For example, in the first two of the three questions we just worked on we saw the mapping $1 \longrightarrow 3$ first taking 0.5 to F, then A to C.

"Are there any other pairs of designs that share a reproduction-action? How can we find all the enlargements realized in our collection of figures?"

Development: The students just sit down and start calculating random enlargements. The teacher holds out for a system that represents each and every reproduction. She puts a grid on the board with all of the images designated down the side as models and across the top as images, and gets the students to fill it in. This can be a skill exercise, or an effort of a small group armed with a calculator, or a little competition: "Who can find the smallest? The largest? One between this number and that?"

The formulations are not simple, but the children manage to master them, and to laugh at the apparent contradictions that they produce.[4] They finish by putting them all in order, from smallest to largest and checking out the effects of various of them.

---

[4]"The more you pedal less hard, the less you go forward", as a child once explained to a flabbergasted psychologist.

> **Remark** This lesson can be omitted for fifth graders, but it demonstrates very nicely the need to distinguish between the mapping that produces the reproduction and the image that it produces. Students can get by with thinking of enlargements as operations or the result of operations without being required to make a formal distinction, but the moment the problems start getting complicated, the teacher is left without any way to explain things to the students who are the least competent at constructing their own models. Teachers then have recourse either to teaching algorithms (the traditional solution) or waiting until the questions can be presented formally (current solution). In either case, there is no negotiation and no teaching of the meaning. The difficulty is not resolved, it is just disguised.

Lesson 6: Reciprocal Mappings

Presentation of the Problem

"When we took M as a model, we found that the enlargement $1 \longrightarrow 1.25$ produced E as a copy. Now we want to know what would happen if we took E as a model and M as the copy. Every length on E corresponds to a length on M. Is it a good (proportional) reproduction? And if so what is the enlargement factor?"

Protest from the class: "It can't be an enlargement! It's a reduction!"

Assignment: "Since you are sure it's a reduction, let's find it!"

Development: This proceeds via an exchange of remarks, propositions and objections between the teacher and the children (and among the children themselves.)

The teacher writes up the beginnings of a table, with E and M at the top, and the children immediately propose to put in the corresponding measurements, starting with what 1 (in E) maps to in M. This one they calculate very swiftly, and find that the mapping in question is $1 \longrightarrow 0.8$. But they still have to verify that this reduction stays the same for all the measurements. This they do individually, though a lot of them think it's unnecessary. Why? "It's just gotta be!" – but they can't articulate a reason. The teacher refrains from making objections.

Information from the teacher:

"The mapping that takes E to M is the mapping reciprocal to the one that takes M to E. (She writes "reciprocal mapping" on the board.) Do you think that every proportional reproduction that we have seen has a reciprocal? If so, would you know how to calculate it? They will also be proportional reproductions."

Exercise: What is the mapping reciprocal to $1 \longrightarrow 5/4$?

Some of the students have to re-do the tables and calculations. The result is either that the reciprocal mapping is $1 \longrightarrow 4/5$ or $1 \longrightarrow 0.8$, depending which tactic the student used.

Challenge: "See if you can find a mapping that is equal to its own reciprocal."

Results: This activity is relatively simple for all the children. It develops rapidly as a game (question and answer.)

Module 10: Multiplication by a Rational Number

Lesson 1: Multiplication by a Rational Number

The process starts with a review of everything the class knows about fractions, bringing back into focus the original construction of fractions as a measurement.

"We constructed fractions, what did we do with them?"
"We put them in order"
"We added them,"
"We did some subtraction problems"
"We converted them into decimal numbers"
"What else do you think we could try to do with them?"
"Multiply them!"

> "We have already calculated the products of two fractions, but we didn't recognize it. We did some calculations that we could have written as one fraction times another fraction. We are going to see if we can find which calculations they were.
>
> We'll need to figure out what it is that lets us put the x sign between two fractions. Why do we have the right to write that when it's a different multiplication from the one we know?"

Remark To justify the use of the + sign on fractions the students contented themselves with verifying that the material operation they carried out, on lengths, for instance, corresponded well with what they were in the habit of associating with addition.

Here the meaning of the product of two fractions is quite different from that of the product of two natural numbers. The only really legitimate way to accept the sign "multiply" would be a detailed examination of the formal properties of the new operation and comparison with the known properties of multiplication. We think that such an exhaustive examination is inappropriate with children of this age, but that it is indispensable to have them inventory a certain number of properties

Either that are conserved (for example distributivity over addition),
Or that change (for example the fact that the product of two whole numbers is equal to or greater than each of the two)
And, of course, to construct a new meaning for multiplication.

Definition of the product of two fractions.
"We know that $3 \times 2/5$ is $2/5 + 2/5 + 2/5$, but is there an addition problem that could replace the operation in $3/7 \times 2/5$? As you might suspect, we need to look at enlargements and reductions and not at additions to construct this new multiplication. We will proceed in three steps.

First step: Let's see if we can find an enlargement in which we might be led to write $3/7 \times 2/5$"

The teacher gives the students a few minutes to think about it and possibly write something on a small piece of paper and put it on the corner of her desk.

At the end of this first short period of reflection (2 or 3 min) the teacher doesn't ask the students for their answer. They will find out whether they were right in the course of the class research.

"Do you know an enlargement that we could call 'x 4'? Up to now we haven't put a × sign in front of numbers that designate an enlargement. But people often do put that × sign. See if you can understand why."

The expected response is: "1 on the model corresponds to 4 on the reproduction" As soon as she gets it, she writes it on the board as a conclusion:

Model   Reproduction
1   ─────────────────────→   4

"Why can we call this enlargement x 4?" And he adds the following measurements:

Model   Reproduction
1   ─────────────────────→   4
5   ─────────────────────→
3.5   ─────────────────────→
3/5   ─────────────────────→

A student comes to the board and fills in the list of reproductions with 5×4, 3.5×4 and $3/5 \times 4$, and often gives the result of the multiplication.

"We can call this enlargement "× 4" because the image of a number is calculated by multiplying the number by 4. Would you know the same way what an enlargement × 5 or × 7 would be?"

Second reflection step: "Now are there some of you who can write in their notebook how they might wind up writing 3/7 x 2/5 in the course of an enlargement?"

The teacher lets them think a few minutes, but doesn't call for the answers. If some of the children, sure that they have found it, put on too much pressure, the teacher can invite them to "deposit" their answer on another little piece of paper on another corner of her desk (so that both they and the teacher can know at the end of the process at what point they knew how to define the product.) At the end of this second period of reflection, the teacher poses a new question:

"On the same principle, what would be an enlargement that we could call 'x 0.25'?" She follows the same procedure as for the fraction.

At the end she introduces a new notation:

"We will write the name of the enlargement on top of the arrow, like this:"

Third step:
"Now can you find the circumstances in which we could write 3/7 × 2/5?"

And she waits until they give the answer

"1 on the model corresponds to 2/5 on the reproduction
3/7 on the model corresponds to 3/7 x 2/5 on the reproduction."

She can either take answers from the students if she thinks enough of them have it right or give one more piece of information and ask: "If we use the enlargement "× 2/7", what are the images of the following?

Model   Reproduction
                     × 2/7
1   ————————————————————————→
2   ————————————————————————→
5   ————————————————————————→

Is it an enlargement or a reduction?" Students write their answers in their notebooks.

Fourth step: Calculating the product of two fractions.

Assignment: "Now can you find what we can do to decide what $3/7 \times 2/5$ means? Do you know how to calculate the result?"

                    × 2/5
1   ————————————————————————→   2/5
                    × 2/5
3/7 ————————————————————————→   [ ]

The students work in pairs. They try to reactivate the techniques they discovered in the activity from Module 8, Lesson 3.

Strategies observed:
The students calculate by using either 7 or 3 as an in-between number, and get (in either case) the image 6/35.

Comment: Every year some children write directly

$$3/7 \times 2/5 = 6/35.$$

The teacher, who goes from group to group during the working phase, expresses astonishment and asks how they got it. They tell her they multiplied the two numerators and the two denominators. "Why?" "It's just what you're supposed to do!"
The teacher tells them that she can't accept a step if they can't prove it's right. The children then solve it with one or the other of the strategies above.

Collective correction: The children who found it come to the board to demonstrate their strategies. It's a rapid reminder since they did these calculations many times in the course of the previous module. Many observe and affirm that you can do it by multiplying numerators and then denominators.

The teacher agrees to check out this method, which she baptizes with the name of the student who proposed it.

Study of the method: $\dfrac{\text{Product of numerators}}{\text{Product of denominators}}$

"Do you think this rule always works, no matter what the fractions are?"

The children hesitate and ask the teacher to give them another product. So she gives them $5/7 \times 4/3$. The children set out to calculate it both ways, and the teacher helps the ones who are having a little trouble. In the end they discover with pleasure that the rule works for this one, too, and enthusiastically endorse the "rule" because it worked again.

But the teacher points out that they only tried two examples and if they are going to adopt it ("institutionalize" it) it has to work *all* the time, on any pair of fractions. So she proposes a new form of verification in the form of a game.

Verification of the rule

First game:

1. The teacher asks the children to choose a number corresponding to these letters: a = , b= , c= , d= . Each child writes on scratch paper.
2. Calculate $(a+b)$: "Are you all going to get the same thing? Why?"
3. Calculate $(c+d)$: "Are you all going to get the same thing?"
   "No!"
4. Calculate $(a+b)+(b+c)$
   Then $a+b+c$
   Then $(a+b)+(b+c)-b$

What do you notice?
The result is written on the board:

$$(a+b)+(b+c)-b = a+b+c$$

Why?

Remark: The children love this activity. Most of them have seen older children calculating with letters and they say so: "It's just like 6th grade!"

Second game: "What does $a/b \times c/d$ mean?

The teacher has a student come to the board while the others look on and comment. The students sets up the usual format

$$1 \xrightarrow{\quad \times\, c/d \quad}$$

$$a/b \xrightarrow{\quad \times\, c/d \quad}$$

With a little help from the teacher, the student works through the whole pattern, getting $(a\times c)/(b\times d)$

Conclusion: The teacher points out that just the way it was in the first game, the letters could represent any numbers at all, so the rule they discovered holds true.

Lesson 2: Multiplying by a Decimal (Summary of Lessons)

The next day's lesson repeats roughly the same process as Lesson 10-1, but with decimal numbers: a search for an example giving rise to 1.25×3.5 and calculations. Clearly it's a question of calculating the length of the image of 1.25 in an enlargement that takes 1 to 3.5. The calculation is carried out initially by expressing the decimal numbers as decimal fractions, then directly, after a rediscovery of the way moving the decimal point represents the denominator. The algorithm is recognized and practiced and given the status of something to be memorized, as it would be in the classical methods.

Results: All the children understand the algorithm and the meaning of this multiplication. But it is interesting to note that it gives them great satisfaction to be able at last to multiply two decimal numbers.

In fact, they have invariably long since been asking the teacher, "Why aren't we learning yet how to multiply two decimal numbers, because we would know how to do it!" The pressure is particularly heavy in the course of the activity about the Optimist (Module 8, lesson 7) because at that point there are always one or two students (either repeating the class or coming in from other schools) who calculate the images of the measurements in the Optimist (whole or decimal numbers) directly by multiplying them by the enlargement or reduction factors.

Since the teacher doesn't take this procedure into account, and doesn't exhibit it when the class does its collective correction these children feel ill treated and ask why their solution hasn't been considered. Often there is one whose response to the teacher's "Because we haven't learned multiplication of two decimals and you don't know what it means" is that he does know – he has learned it.

In that case, the teacher has no choice but to have a collective clearing-up session. She reminds the students of the meaning of the different multiplications that they have already dealt with:

4×125 means $125 + 1,125 + 125 + 125$

4×2.5 (where 2.5 may be the length of a stick or the price of an object or the capacity of a container) means $2.5 + 2.5 + 2.5 + 2.5$.

But what does 1.7×0.94 or 4.128×3.67 mean?

Obviously, the children then realize that there exist multiplications whose meaning is different from those that they know, and they all agree that these calculations can't be used at this point in the progression

It's easy to understand the relief they feel and express during this session, and their desire to do and use this long-awaited calculation!

Lesson 3: Methods of Solving Linear Problems (Summary of Lessons)

Introduction (for the teacher) The examples that follow will permit us to describe the typical progression of the study of a problem and to indicate how the teacher and students draw conclusions from it that are explicit but definitely not learned by heart. We will also indicate as many as possible of the conclusions and remarks that the children may make as they master different methods of solution, different types of questions, different uses of linear functions, etc. In order to avoid presenting a multitude of problems we will concentrate all these conclusions slightly artificially on a few examples. In any case, from the moment that the students start solving the problems the teacher ceases to exercise control over the details of the means of coming to the conclusions, and focuses on keeping the class engaged and with its eye on the goal.

The teacher demonstrates as an example what questions it is worthwhile asking oneself in the course of solving the following problem, while explaining step by step the solution of the problem:

The children collected the cream from 2 l of whole milk and got 32 cl of cream. They also collected the cream from 5 l of low-fat milk and got 40 cl of cream. Can you answer the following questions:

How much cream would you get from 50 l of milk? 125 l of milk? 250 l of milk? How much milk would you need to get 4 l of cream? 2 l of cream? 10 l of cream?

The students discuss the problem and end up asking the teacher to remove the ambiguity of the questions. This helps prepare them to construct problems themselves.

The teacher makes comments and indicates how to present the givens, how to express the results (in the solution) and how to check the use of numbers and functions. Then he asks the children to recall the different ways they have encountered to solve linear mappings.

Lesson 4: The Search for Linear Situations
(Summary of Lessons)

In this lesson the teacher sets up a "tournament of problems ". On a regular basis, students are to come up with problems that involve solving a linear mapping. The problems can be invented or taken from a book, but in any case the student who presents a problem must be able to give a solution if asked.

The tournament will be open until the end of the year. From time to time the class will spend a few minutes "judging the problems" the way pictures are judged at a painting exhibition: which is the most interesting, the most beautiful, the most original, the most trivial – but it is not the students who are being judged, it is the problems. Only the teacher knows which student proposed which problem.

The primary goal of this activity is obviously technical: it develops in the students a knowledge and culture of problems. By trying to classify them: problems about sales, about representations (in the ordinary sense), about relations between physical sizes, about percentages, etc., they observe their similarities and differences and varied characters. They will know them much better than if they were using the teacher's choices. This is not the place for a standardized classification!

It's the activity that matters more than its result. And the traps and counterexamples stand out without it being necessary for a certain number of students to fall victim to them. The discussions, of course, point up ways to look for more examples.

The second goal is in effect psychological. This set up provides a nice safe zoo where they can approach the wild beasts to which they often fall victim.

Module 11: The Study of Linear Situations in "Everyday Life"

Remark to the teachers: In this initial paragraph, we will first familiarize the children with the designation of linear mappings using the vocabulary of fractions.

The problems and the examples should thus be chosen appropriately. For example, in the problem:

"The 25 students in a fourth grade class go to the swimming pool every week. At the end of the year, 4/5 of the students know how to swim. How many students is that?"

The fraction 4/5 is a **ratio** between two sizes: the number of students knowing how to swim and the total number of students. It does not correspond to a **linear mapping**: we are not talking here about a rule for determining how many swimmers some other class should have. In fact, one might expect to compare the ratios. On the other hand, problems that correspond to "rules" – conventions or logical necessities – do furnish examples of linear mappings.

Examples of "rules":

Composition of a food product (milk, bread, coffee, etc.) and transformations of it: coffee loses 1/7 of its weight in roasting; making fig jam requires a weight of sugar equal to 3/4 of the weight of the fruit, etc.

Lesson 1: Fraction of a Magnitude

(a) *Introduction*: Assignment

Weight of fruit in kilograms		Weight of sugar in kilograms
2	⟶	1.5
12	⟶	9
8	⟶	6
5	⟶	3.75

Is this table produced by a linear mapping?

(b) *Development*:

Students verify it by the methods available to them:

Is the weight of sugar corresponding to the sum of two weights of fruit equal to the sum of the weights of sugar corresponding to the two weights of fruit?

First method: Some students suggest comparing the weight of sugar (15 kg) corresponding to $12+8$ kg of figs with that corresponding to 4×5 kg of figs ($4 \times 3.75 = 15$), but they are already assuming that the mapping is linear. Here this method provides no verification.

Second method: Is it true that if we multiply each weight of fruit by some particular number we will have to multiply the corresponding weight of sugar by the same number to find the new amount of sugar?

The children verify that $2 \times 6 = 12$, does indeed correspond to $1.5 \times 6 = 9$, and $2 \times 4 = 8$ to $1.5 \times 4 = 6$. Then they realize that they are going to have to do an awful lot of verifications (16, or at least 6). So some suggest working out 1 and using it to verify the rest.

Third method: Do you get the weight of the sugar by multiplying the weight of the fruit by a constant?

To find this number, the children look for the image of 1, then carry out the multiplications 12×0.75; 8×0.75, etc.

Conclusion: the mapping is linear.

Note: The children already know these different methods, and use whichever is the most efficient in a given situation. They will be inventoried and institutionalized a little later.

(c) Summaries of the table

"How can we summarize this table in a short recipe?"
The students propose their standard method using the image of 1:

"You have to multiply by 0.75", swiftly corrected to
"You have to multiply the weight of the fruit by 0.75 to find the corresponding weight of sugar."

The teacher has them convert to fractional notation and simplify the fraction to 3/4.
$75/100 = 150/200 = 15/20 = 3/4$
Then he reformulates it as

"You have to multiply the weight of the fruit by 3/4 to find the corresponding weight of the sugar."
"You have to apply x 3/4 to the weight of the fruit to find the corresponding weight of the sugar."

He tells the class: "You will often find this said with expressions like
The weight of the sugar is 3/4 the weight of the fruit.
To find the weight of the sugar, you take 3/4 of the weight of the fruit, or you calculate 3/4 …
What you have just done here is to **calculate a fraction of a number.**

Notice that in the table we find opposite the number 4 (in the weights of the fruit) the number 3 (in the weights of the sugar). The weight of the sugar is 3 when the weight of the fruit is 4; the ratio of the weight of sugar to the weight of fruit is 3–4.

Careful! The ratio of the weight of fruit to the weight of sugar isn't the same! What is it?"

Exercises in Formulating Fractions in Terms of Linear Mappings

(a) *Assignment*: Here are some situations formulated in this way.
    Translate them into the linear mapping schema.
    Then pose some questions, if necessary filling in needed information.

1. A merchant wants her profit to be 2/5 of her purchase price.
2. Wheat gives 4/5 of its weight in white flour.
3. Draw a rectangle whose width is 2/3 of its length.
4. To buy on credit at a store, you have to deposit 3/8 of the selling price at the time of purchase.

(b) *Development*: Recognition of the mapping designated by a fraction and search for a schema.

The children know for example that the first situation has to do with a × 2/5 mapping, and they can represent it by

$$1 \xrightarrow{\times 2/5} 2/5$$

For them, the problem is to know where to put the price and the profit.

Often, if the reference situation is well known, outside information comes in to indicate the solution, for example, in the case where one is taking a part of a whole. Example (not true for this particular case) an image that is smaller than the original quantity… if you take two fifths, then 5 can't correspond to 2, so 2 must correspond to 5.

Here the "semantic" information was intentionally rendered inoperative. The merchant could wish to make a profit of 5/2 of her purchase price, because the profit is not part of the purchase price. Also the situation is not well known to the children. In this case, the formulation itself must be consulted: the expression "2/5 **of** the purchase price" shows that 2/5 **is not** the purchase price – that the purchase price is what you're taking 2/5 **of.**

So we have

Purchase price
$$1 \xrightarrow{\times 2/5} 2/5$$

and it must be that what we get to is the profit.

Here the students go back over the formulations they have already encountered:

For a purchase price of 1, the profit is 2/5
When the purchase is 5 the profit is 2 means that the profit is 2 for [a purchase of] 5
The profit is [2 per 5] (purchase price)

Two fifth is the profit; you have to multiply it by 5 to get 2 times the purchase price:

$$
\begin{array}{ccc}
1 & \xrightarrow{\hspace{2cm}} & 2/5 \\
\times 5 \downarrow & & \downarrow \times 5 \\
5 & \xrightarrow{\hspace{2cm}} & 2
\end{array}
$$

In response, students produce schemas such as

Purchase price $\xrightarrow{\times 2/5}$ profit

Weight of wheat $\xrightarrow{\times 4/5}$ Weight of white flour

Sale price $\xrightarrow{\times 3/8}$ deposit

Length of rectangle $\xrightarrow{\times 2/3}$ Width of rectangle

Remarks for the teacher:

1. The starting number multiplied by the number determining the mapping is equal to the ending number. Since we can get the starting number by dividing the ending number by the fraction, distinguishing the starting and ending numbers is closely linked to the understanding of the product of two fractions or of two numbers.
2. The children should interpret the formulations directly. They must not be formally taught "algorithms" for getting the right answer. Numerous exercises and translations among the different formulations, accompanied by arguments of every sort (most of them particular to a specific example and thus not generalizable) will enable them to make sense of the cultural formulations that they will run into (of which a few are pretty illogical.) In any case, this study will be resumed in Module 14, where, with the composition of mappings, it will be possible to get back the traditional meaning (3/4 means divide by 4 and multiply by 3.)

(c) *Development (continued):* Formulation of questions and problems; search for necessary complementary information.

The students propose, for each of the above situations, several problems obtained by adding questions. For example (for the second situation), they might ask

- The weight of flour that you get
- The weight of grain that was necessary.

But this then requires that one know the weight of the grain in the first case and the weight of the flour in the second.

The children have no difficulty in posing these questions thanks to their familiarity with the tables. Nonetheless, this activity brings up interesting remarks on the relevance of information and questions.

Example: "A car has gone 100 kilometers and its tank is 3/4 empty."

If we add that the tank was full at the start, then we can ask how many more kilometers it can go:

3/4 of a tank $\xrightarrow{\hspace{3cm}}$ 100 kilometers

1/4 of a tank $\xrightarrow{\hspace{3cm}}$ (100 ÷ 3) kilometers

Since 1/4 of a tank is left, it can go 33 km.

But if instead of the information that the tank was full we say that 20 liters of gas are left, then what we can ask is the capacity of the tank:

1/4 of a tank $\longrightarrow$ 20 liters

4/4 or a full tank $\longrightarrow$ $20 \times 4 = 80$ liters

(and in this case the number of kilometers is useless.)

Further remarks for the teacher:

(i) Students often have difficulties in identifying the three elements of the mapping:

    (a) The domain set – the thing you are "taking a fraction of", which is at the front of the schema with the arrows, but often named after the fraction is named, as it is in all of the examples above, and at times difficult to identify.

    Examples:

    "The reservoir is 3/4 empty"
    "A worker earns a certain amount and saves 3/40 of it."

    (b) The correspondence: the way of finding the image of a given number. Classically, 3/4 describes the operation "multiply by 3 and then divide by four" (or possibly vice versa), which we will describe in module 12. Here the student says that 1 corresponds to 3/4, without making reference to some operation for getting from the 1 to the 3/4. This way we avoid various difficulties linked to

    – The impossibility of actually making the division being envisaged
    – Too concrete a representation (taking fourths of a sum!)
    – Or the complexity of the concrete operations envisaged.

    But this makes the a priori identification of the image set all the more vital
    (c) The image set of the values that are "a fraction" of another one is at times all the more difficult to distinguish in that the French language[5] permits a constant confusion between an operation and its result, a mapping and its image, an action or fact and the state that is its consequence (the marriage took place on such and such a date and lasted 20 years!)
    The language of fractions assumes that it is obvious how to carry out a linear mapping.

(ii) We assume that the schema can be made by the students they before know the question posed and independently of it, using only the language of the representation of the situation of reference. We assume next that the question can be represented before and independently of the solution to be produced.

_____

[5] As well as the English one!

(iii) The students are thus invited to pose questions and to inventory the possible questions:

- The search for the image (a fraction of a quantity)
- The search for the object (the quantity of which we know a fraction)
- The search for the mapping (the fraction taken – or the ratio of the two magnitudes)

Questions provide the justification for the rest of the lesson.

(iv) The students comment that they "can't calculate anything" if they don't know the quantity or the number of which they are "taking a fraction", but they can make the table, just as they could draw a rectangle whose width is two thirds of its length.

Calculations with "fraction-mappings"

(a) *Calculating the image*: The teacher presents the following problems:

1. You buy 6 kg of fruit to make jam. This kind of fruit gives 2/3 of its weight in juice. You need to add a weight of sugar equal to the weight of the juice. How much sugar should you buy?
2. Cotton shrinks when it is washed: it loses 2/9 of its length. If a piece of cotton fabric measures 6.75 m, how long will it be after washing?
3. 4/25 of the volume of milk is cream. How much cream would you get from 3/4 of a liter of milk?

(b) *Mathematization with the children*:
The teacher invites the children to make mathematical remarks about the problems they have just done. Some of them comment that they have calculated a fraction of a whole number, then a fraction of a decimal number, and finally a fraction of a fraction.

The teacher requests more precision: "What operation did you do to take 2/3 of a number?"

"We multiplied the number by 2/3."
"So what operation can we use to describe the mapping 'take 2/3'?"

By similarity to Activity 10.1 (product of two fractions), the children propose:

$$1 \longrightarrow 2/3 \quad \text{is multiplication by 2/3!}$$

$$1 \xrightarrow{\times 2/3} 2/3$$

The teacher then puts a frame around

Taking a fraction of something means multiplying by that fraction.

(c) Calculating a number of which a fraction is known.
For a holiday dinner, you buy a 3.6 kg roast. When it is deboned and cooked, this meat loses 2/9 of its weight.

If you want to have to have 2.1 kg of meat after cooking, how many kilograms do you need to ask the butcher for?

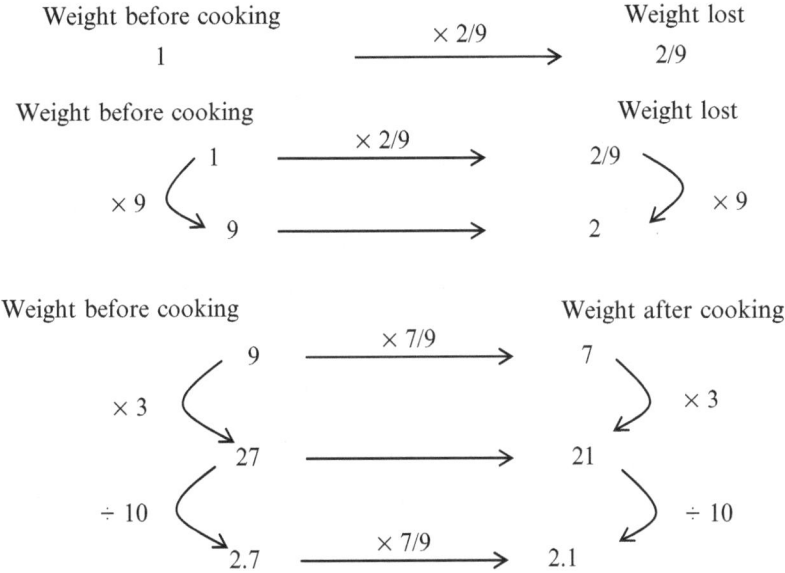

The teacher asks what operation corresponds to this mapping:

9 ⟶ 7

1 ⟶ 7/9

⟶ × 7/9

Summary of the Remaining Paragraphs and Sections of Module 11

The rest of Lesson 1 takes up the reformulation in terms of fractions of the mappings that they already know: enlargements, etc. Simplification of fractions is recognized as the search for smallest whole number correspondences. The students study different ways of realizing certain fractions of squares or of dividing a segment into any number of parts using a supply of equidistant parallel lines.

Lesson 2 is dedicated to the study of percentages. Lesson 3 presents the correspondences of the same magnitudes (scales). The determination of the length of a segment that is out of their reach leads the students to surround it with a rigid figure and make a reduced model of it. The issue is to show a simple use of scales. By studying the correspondence between different magnitudes, the teacher leads the students to the use of non-scalar linear coefficients like the price per meter or the pounds per unit of volume.

While the second two modules in this section will provide students with occasions for addressing the often hidden difficulty of distinguishing between problems involving ratios and those involving linear mappings, at this stage the original manual simply presents the teacher with a warning of that difficulty.

The Problem-Statement Contest (Commentary 2008)

The curriculum on which we ran our experiments from 1973 to 1980 introduced different aspects of rational and decimal numbers by following new mathematical pathways. It consisted of the modules 1–11 and module 14. Teachers completed it with a classical exploration of the official program – the metric system and application problems – with the standard organization. We were interested, among other research subjects, in evaluating the impact of our new introduction on the performance of the students in these well known domains. Starting in the late 1970s the study of the *didactical contract* led us to study problem solving.

The contest of problem statements, introduced in Module 10.4 of the manual, recounts the first endeavors of this program, which continues to be carried out today.

Developing on the ideas of Polya, who suggested making students experts in the solution of problems by teaching them methods and heuristics, we wanted the students to stop thinking of problems as tests or as individual challenges and instead to develop a culture made up not just of tasks observed, techniques, and known results, but also of knowledge under construction and of emotions. Heuristics can be a useful tool, but if they are taught as skills to learn and apply, they become nothing but bad theorems.[6] Analogy is likewise a helpful but flawed tool: we have shown, among other things, how the abuse of "analogy" raised to the rank of a teaching principle resulted in an augmentation of students' failures.

(continued)

---

[6] We have shown that if the teaching of "problem-solving methods" follows the classic conceptions relative to knowledge and learning, the teaching will lead to uncontrollable metadidactical slippage and to failure. [Metadidactical slippage is discussed in Chap. 5.]

(continued)

The object of the ***Contest of Problem Statements*** is to improve students' learning by leading them to consider classes of problems and solutions and not just types of problems, to examine what constitutes them, to discuss them. Above all, it is to change the psychological, didactical and social conditions of the students' activity by changing their position in the didactical situation.

Instead of the usual pattern of using the problems to judge the students, here the students judge and determine the value of the problems. Thus, as is done for works of art at a showing, the students themselves award prizes among the problems presented to them: a prize for the longest, the most interesting, the most difficult, the most annoying, the most original, etc.

Ordinarily, problem statements are presented by the teacher. In our project, the students choose some and above all produce some themselves and discuss them with each other and with the teacher.

For teachers and society, problem solutions are regarded as a source of knowledge about the students. Here the solutions become knowledge of the student about mathematics and about the problems.

Thus the search for and making of "original" problems can become a challenge for the students, a motive and instrument for recognition, comparison and classification of problem statements, on condition, clearly, that the focus remain their mathematical solution.

The most important point concerns the status of the *connaissances* that appear and are formulated by the students or the teacher *in the Situation,* and their evolution toward the status of *savoirs.*[7]

In this process, the students learn to analyze these texts, to pose questions, to distinguish the givens. Statement and solution form, in fact, definitions and theorems. The teacher must thus know their grammatical and logical components. It is essential that he not teach them as *savoirs* and especially that he not explain them. The teacher's situation is similar to that of parents of a young child who is learning her language, in that the parents must get her to respect phonetic rules without explaining them. The classifications the children come up with are *connaissances* not *savoirs.* To take this knowledge as a piece of *savoir* constitutes a metadidactical slippage (teaching the meta-object in place of the object) which generally leads to other slippages.

---

[7] *Connaissances* and *savoir* both translate to "knowledge", but they are used very distinctly. A good working definition is that *connaissance* is general knowledge and *savoir* is reference knowledge. For a more nuanced definition, see Chap. 5.

Module 12: More on the Problem Statement Contest

Lesson 1

Phase 1: Research

(a) *Assignment*: "We will continue our problem statement contest (cf. Module 10, Lesson 4).

Today you are to propose problems that lead to doing a division. You will write the problem statement and simply carry out the operation, but you should prepare the justification for the operation in your head so that you can give it orally to your classmates. You may start with simple examples or ones we have encountered before. Don't get complicated – introduce the dividend and divisor with a sentence and ask for the quotient. What we are interested in is the occasions for doing division.

On the other hand, do try to put as many decimal numbers or improper fractions as you can in the statements and the solutions PROVIDED THE PROBLEM STILL MAKES SENSE.

The session continues. Students present their problem statements and verify under the teacher's guidance that the statement is correct (givens and question) and plausible, and that the solution offered is right. They inventory a certain number of difficulties: confusion between the means of calculating (example: division) and the statement or the means of checking (example: multiplication), classification by a partial calculation, etc.

Phase 4: Production of New Problems and Use of the Criteria

(a) **Examples: creation of a category**

The students propose to put the following two problems into the same category:

"Evelyn divides a 1.50 meter long ribbon into two equal parts. How long is each piece?", and

"Three brothers share a sum of 375 francs equally. How much does each one get?"

T: "Why do you think they are alike?"
S: "Because something is being divided in equal parts."
T: "Still, there are some differences?"
S: "Yes, in this one it's ribbon and in that one it's money."
T: "And can you divide up ribbon the same way as money?"
S: "??? No, but for numbers it's the same."
T: "Find an 'intermediate' problem that shows the similarity: for example, replace the givens from one with the givens from the other one."

The students propose the problems:

"Three brothers share a sum of 1.50 francs equally. How much does each one get?" and
"Evelyn divides a 375 meter long ribbon in 3 equal parts. How long is each piece?"

The teacher accepts the similarity provisionally – this will be a category of "divisions" – but brings the number problem back up.

(b) **Creation of criteria**

T: "So, changing the numbers doesn't change the problem?"

The students are ready to think that changing the numbers does not lead to changing the operation.

T: "Let's take the same problem statement and change the numbers. What's going to happen?"

The remarks that follow can make it possible to clarify the effect of the magnitude of the numbers:

Some numbers are plausible and others are not (value of the givens): a 375 m ribbon is unusual, but a 375 km ribbon is impossible.
If Evelyn divides up a 1.523712 m long ribbon the problem is plausible, but the precision is ridiculous (representation of the givens.)
If the number of brothers were the decimal number 3.2, the problem would make no sense (nature of the numbers.)

The Classifications of Problems (Commentary 2008)
We classified problems about division of rationals using the following criteria:

Classification according to the material or symbolic manipulations carried out:
   long division of natural numbers, then decimal or rational numbers, measurements, exchanges, successive approximations, equal or unequal shares,...

Classifications according to special vocabularies
   Arising from practical or professional activities (scales, percentages, rates)
   Arising from applications (speed, physical density, etc.)
   Arising from some cultural vestiges (fractional measurements)

Classification according to problematics[8]

Classification according to mathematical concepts,
   Either classical (types of operations, fractions, ratios, proportions),
   Or more current (order, topology, algebraic laws and structure, measure, scalar, function, ...) which are the criteria maintained in the course.

---

[8] A *problematic* is something that constitutes a problem or an area of difficulty in a particular field of study [Oxford English Dictionary] The French use problematics more specifically to refer the set of questions posed in a science or philosophy with respect to some particular domain.

In Lesson 2, the first problems to classify are those that are familiar to the students. They arise from the conception of long division.

The next paragraph brings up the review of long division based on manipulations: sharing, partitioning, attribution and distribution lead to different strategies according to regular or irregular conditions (leading to the equalization of parts, for instance.) The term "division" unites certain of the conceptions, but not all (such as the search for a remainder.)

Following that, classification according to problematic leads to envisaging the calculation of the unknown term of a product or that of a component of a product measurement. Each conception leads to different manipulations which themselves bring up different modes of calculation. Classical teaching requires that children recognize division "naturally" as the concept common to these varied conditions and at the same time that they support this recognition with concrete arguments!

The extension of long division to division of decimal numbers happens naturally with the method of successive subtractions, which makes it possible to rediscover the reasoning and establish the algorithm for bracketing a number [see Module 5.]

The use and comprehension of division of decimal numbers are facilitated by its similarity to long division in the natural numbers. This is a recognized fact. But it is important to note that this facility hides a difficulty that is easy to observe, which itself hides an epistemological and didactical obstacle that is fundamental for the passage from the use of natural numbers to that of rational and real numbers.

Long division is based on the idea of measuring something using something smaller as the unit. If by considering only the whole number parts of the dividend and the divisor the student can conceive of the long division that solves the problem next door to the problem required, he can simply extend the algorithm by the calculation of decimal parts. For example, $17.4 \div 3.62$ is understood first in the sense of $17 \div 3$, the rest is a matter of the algorithm. This conception collapses if the long division indicated has no meaning. More explicitly, if the divisor is greater than the dividend, or if the dividend is less than two. The operation $0.4 \div 0.62$ is the case that gives rise the most difficulties.

Modules 14 and 15 make it possible to surmount this difficulty in conceptualization by means of a deep comprehension of the structure of rational and decimal numbers.

Module 13: New Division Problems in the Rationals

The first lesson continues the inventory of problems that was undertaken in module 12. The issue is first to have the classification completed by adjoining new problems where the division is defined by a rational linear mapping (i.e., one with a rational coefficient) expressed in any way.

The class should find there the notions studied in the preceding modules, and can make an inventory of them: a measurement divided by a scalar, a measurement divided by a measurement of the same type or of a different type, a scalar divided by a scalar, etc. The class can recall that a division also consists of finding a decimal expression (exact or approximate) for a fraction (module 7). This uses the idea that the result of a division expresses the measurement of the dividend if the divisor is taken as the unit: for example $12 \div 3$ expresses that if we measure 12 with 3 as a unit, the result is 4. Measuring 3 with 4 as a unit gives a result of 3/4, like dividing 3 into 4 parts, ... and if the numbers are measurements in meters, then the result is __ meters.

This leads the students to understand that a fraction is the indication of a division that one neither can nor wants to carry out, but about which one can calculate.

(a) But they may also discover that the mappings sometimes pose difficulties. An example of these difficulties: the students know how to find the result of dividing one fraction by another as long as the first is the measurement of the thickness of a piece of cardboard and the second expresses the thickness of a sheet of the paper that makes up the cardboard. But what does $3/4 \div 2/5$ mean in general, in particular when the result is not a whole number? Interpreting this operation with the general idea that division is partitioning does not furnish a practical procedure. The equivalence of commensuration and partitions of unity always presents difficulties.

(b) Since problems often disguise the distinction – well known to our students – between ratios and linear mappings, the teachers propose problems of the nature of the following example – which does not figure in the 1985 Manual:

"A father is 5 times as old as his son. How old is the son if the father is 35 years old? How old will that father be when his son is 10 years old?" For the students, the issue is to recognize that the ratio between the age of the father and that of his son does not determine a linear function between their ages: the father will not be 50, but only 38. There is indeed a function, which fools the children, but it is a translation (+28). Clearly if the question had been the age of the father when the son is 35 years old the error would have been easier to detect.

(c) The study continues with the inventory of the roles of division in the study of a linear mapping: finding the correspondent (the image) of 1 when the coefficient is known, finding the ratio between the two values, calculating the coefficient of the mapping when one original and its image are known, calculating the original when the mapping (the coefficient) and the image are known, etc. The numbers are decimals or fractions.

At this stage, the students conceive of all linear mappings as multiplications (for example, $\times 3/4$).

Lesson 2 has them study linear mappings that are read as "divisions" and that will be understood as the reciprocal of multiplication by a number.

Lesson 2: (Extract) Division as Reciprocal Mapping of Multiplication (The Term Is Not Taught to the Students)

The session proceeds in the form of a sequence of problems that the students carry out rapidly. These problems provide the occasion for posing some mathematical questions. The teacher needs to make clear the distinction that he makes between these mathematical questions and the problems. The mathematical questions are the real object of challenges proposed to the students, the occasion for debates, and the real goal the teaching is aiming for. Problem statements, whether proposed by the teacher or by the students, are there only as means of treating those mathematical questions, or as applications of knowledge newly acquired or discovered.

Division (by a Number), a Linear Mapping

(i) **First problem statement:** "A movie ticket costs 35 francs. The total receipts of a theater are:

3,325 F on Monday	5,250 F on Friday
4,480 F on Wednesday	6,125 F on Saturday
3,675 F on Thursday	6,230 F on Sunday"

First question: How many paying customers were there on each of the days of the week?

The students make a table in which they place the results of their divisions by 35. The teacher asks if it is the result of a linear mapping. Students: "If we add up the receipts and divide that by 35 we ought to get the sum of the numbers of tickets" "If there is twice as much money it is because there are twice as many customers."

Division by a Fraction: Calculation of the Image

The issue is to find out how to divide by 3/8. To support their reasoning, the students must think of a problem. For example:

"I have to divide by 3/8 if I am looking for the number of 3/8 mm sheets it takes to make different given thicknesses, for example 6 mm, 9 mm, etc.

(a) Reasoning by an assumption contrary to fact:

If the sheets were 1/8 mm thick, we would need eight sheets to make a thickness of 1 mm, so we would need 48 sheets to make 6 mm. But the sheets are $3 \times 1/8$ mm, so these are three times thicker, so we need three times less (than 48) to make a 6 mm cardboard. So $6 \div 3/8 = 48/3 = 16$.

(b) Reasoning by equivalence:

6 mm is 48/8 mm, so the reasoning above produces $48/8 \div 3/8 = 48 \div 3 = 16$.

(c) Using the reciprocal:

The reasoning that leads to a search for the unknown term of a product leads to sentences like: "If I had 3 sheets of 3/8 mm, they would have a thickness of 9/8 mm – there have to be more. 10 sheets $\rightarrow$ 30/8 – that's a little more than 3 mm – we need still more. It's 16, because $16 \times 3/8 = 48/8 = 6$ mm", which the teacher translates: "So we have to look for the number that you can multiply by 3/8 and get 6. ___ $\times 3/8 = 6$" and he draws the schema

$$6 \quad \xrightarrow{\div\ 3/8} \quad \xleftarrow[\times\ 3/8]{} \quad \boxed{} \quad ?$$

With, admittedly, some effort, the teacher can then obtain a recollection of module 9.6: to find the object, you have to find the image of 6 by the reciprocal of $\times 3/8$.

The reciprocal of $\times 3/8$ is $\times 8/3$, so

$$6 \quad \xrightarrow{\times\ 8/3} \quad \xleftarrow[\times\ 3/8]{} \quad 48/3 = 16$$

From these three methods, one can retain that each time, the student has multiplied the numbers whose image he wanted by 8, then divided by 3.

Division by a fraction is the reciprocal of multiplication by that same fraction (2008 Commentary)

Continuing studies of this nature with other examples was tried, but it is clear that this ambition is not very practicable without a veritable teaching of rules, without intense training with lots of exercises. Even if certain students are able to answer questions of this kind once – and most cannot – the reasoning is uncertain and painful. We explained this difficulty by the complexity and variety of material operations that concretize them and are necessary for verifying them. The solution of this problem is the object of the two modules that follow.

(continued)

(continued)

> To illustrate the difficulty of stating general principles, we report the following observation:
>
> The students noticed that a multiplication a×b=c could give rise to two divisions: c÷a and c÷b. But when one of the numbers is a measurement and the other is a scalar (a ratio or a coefficient) one of these divisions may not correspond to a mapping, especially not a familiar mapping.
>
> *Example*: rate x principal=interest. The mapping principal → interest (rate fixed) and its reciprocal are more frequently envisaged than the mapping rate → interest (principal fixed). At the end of this activity, a lot of the students can answer the question: "One student says 'the reciprocal of (× 7/13) is (÷ 7/13)'. Another one says 'No! The reciprocal of (× 7/13) is (× 13/7)!' Which one is right?" But very few can "prove" it with a calculation.

Lesson 3 brings up various methods of getting the students to consider proportional mappings. A "portrait game" trains the students to recognize and characterize them. A display – the "tapestry of proofs" – helps the students during collective discussions. It enables them to follow and to determine, at any moment of a debate, who claims what and who is supposed to prove what. The aim is to help the teacher to lead the class progressively to distinguish logical argumentation from purely rhetorical exchanges. This design, which was too complex and had insufficient a priori study, never led to any lesson projects that were satisfactory enough to be realized.

The students, with the teacher, pull together and summarize what they now know.

Extracts from the Original Text

Remark: Not every student can achieve a level of comprehension of these questions sufficient to be able to produce individually the proofs sketched below. The proofs should not be required as skills to acquire. Furthermore, the "rules" proposed by the students should not immediately be institutionalized. They should remain in doubt – that is, something to be verified, either by a calculation that uses the representations used in the proofs, which might not be general, or by previously established results.

Conclusions drawn with the students:

1. "We are going to make an inventory of the different ways of writing a linear mapping and of writing its reciprocal.

We have seen that we do the same thing no matter what the numbers are, so let's choose some numbers to work with.

The linear mapping is given by an ordered pair:

$$14 \longrightarrow 27$$

If we want to express it as a multiplication, the mapping is $\times\ 27/14$
The reciprocal mapping is given by the ordered pair,

$$27 \longrightarrow 14 \text{ which is the mapping } \times\ 14/27.$$

If we want to use division to express

$$14 \longrightarrow 27$$

we find the division by looking at the reciprocal expressed as a multiplication. The reciprocal is the mapping $\times\ 14/27$, so the original mapping is $\div\ 14/27$.

Let us present these results in a table, with a and b being two random numbers."

Different ways to designate a linear mapping:

Linear mapping	Reciprocal mapping
a $\longrightarrow$ b	b $\longrightarrow$ a
1 $\longrightarrow$ b/a	1 $\longrightarrow$ a/b
$\times$ b/a	$\times$ a/b
$\div$ a/b	$\div$ b/a

To summarize this table, all we have to remember is that

$$\boxed{\times\ b/a = \div\ a/b}$$

1. Shrinking and enlarging by multiplying and by dividing:

   (a) "We found some mappings that shrank and some that enlarged, all of them expressed as multiplications. Can you give me some?"

   The students talk about this apparent paradox, which they encountered in the lessons on the "Optimist" (Module 9), and which surprised them considerably. At that time they had remarked that "Up to now we thought that multiplying always made things bigger, because the only numbers we knew about were numbers bigger than 1!"

   They propose mappings (which the teacher writes on the board), at the same time classifying them into two categories: those that enlarge and those that shrink (the teacher might add a few, too.)

× 2/5; × 1/2; × 12/7; × 4/4; × 7/5; × 1; × 1/4; × 1.4; × 0.95; × 2.75; …

The ones that enlarge	The ones that shrink
× 12/7; × 7/5; × 2.75; × 1.4;…	× 2/5; × 1/2; × 1.4; × 0.95;…

That leaves × 1 and × 4/4, which call for a reminder from the "Optimist" section: "We saw that if we made a reproduction using × 1 we got the original back!"

Conclusion: the teacher has them explicitly express the conclusion: "The mappings that shrink things are expressed as multiplication by a number less than 1."

(b) By dividing:

"We just recalled that we can shrink a model using a multiplication. Is it possible to enlarge a model using division?"

To increase their comprehension, the teacher suggests that the students find a translation like the one they are in the habit of using in such cases: looking for the image of 1.

$$1 \longrightarrow 7$$

What is the mapping that is expressed as a division and lets us multiply by 7 (or enlarge 7 times)?

The students suggest writing x 7, but that doesn't answer the question that was asked. To help them, the teacher asks, "What can we divide 1 by to get 7?"

$$1 \xrightarrow{\div\,?} 7$$

"It's a number less than 1, because when we divide 1 by it we have to get back to 7. So it must be 1/7."

$$1 \xrightarrow{\div\,1/7} 7$$

The teacher writes: $1 \xrightarrow{:\,?} 5/2$ and then asks, "Can you find some more mappings?" But using the same system, the students only find divisions of the form ÷ 1/n. So the teacher asks them to find a division mapping that takes 1 to 5/2:

As before, they first say "It's × 5/2!" "How can we write it as a division? See if you can find other ways of writing this mapping." The students remember that they had just learned that × 5/2 is the same as ÷ 2/5 (they give the proof: "The reciprocal of × 5/2 is × 2/5, and the reciprocal of × 5/2 is ÷ 5/2. So × 2/5 is the same as ÷ 5/2")

As in the activity before, the teacher has them produce a collection of mappings expressed as divisions and classify them by whether they enlarge or shrink. This they do, observing also that ÷ 4/4 does not change the model.

Conclusion: the teacher has them explicitly express the conclusion: "The mappings that shrink things are expressed as division by a number greater than 1, and those that enlarge are expressed as division by a number less than 1."

Commentary

It might be useful to recall that history has attempted to make of the concept of fraction the universal instrument of measurement and of treatment of proportionality. But in the end this attempt has failed. Today the concept is a mosaic of a plethora of particular expressions in an environment of metaphors that are neither general, nor well adapted to the physical manipulations that they claim to represent, nor well adapted to a general mathematical treatment. An ambition of the reforms of the 1970s was to erase this obstacle a little, but it still holds a major position in our cultures and our practices.

A visit to the field of applications of proportionality traditionally occupied a major part of the program of mathematics. In the 1970s this field was greatly reduced in the curricula of the period. Not wanting to lose any of this essential educational project, we tried to obtain equivalent knowledge with the students, but with fewer lessons specific to different fields and more mathematical reflections, and with a small dose of meta-mathematical and heuristic reflections, on condition that they be formulated by the students and not set up as methods. We would encounter the terms of proportionality on occasion, but we would replace them with the mathematical terms introduced in the lessons.

We knew already that our curriculum (modules 1–11, 14 and 15) brought real improvements to the teachers' and students' possibilities for dealing with applied problems. We also knew already that with the usual conceptions and didactical practices the use of arrows that we had introduced risked provoking a formalist drift and a metadidactical slippage if that use expanded beyond the terrain of the experiment. We then wanted to know what the effect of our mathematical introduction (modules 1–11) would be on problem solving, before the homogenization (the identification of a/b with x a/b.) So we put a first exploration of problems (modules 12 and 13) before the last two modules, which we continued to teach as we had before. Note that this study itself constituted a metadidactical slippage that had to be closely monitored (we expand on this notion in Chap. 5.)

The additional modules were optional. That was a part of the experimental plan. To study the effects of our variations, it was important to maintain the teaching conditions that characterize the whole process under study. It was fundamental to our research plan that rather than the usual practice of evaluating how much

(continued)

(continued)

material it was possible for the students to learn, we set a fixed goal in terms of the content and then compared the efforts and time spent by the teacher and the students to surmount the difficulties within that content. It was thus necessary to let them react to the difficulties, whether by reasonable supplementary effort or possibly by giving up, and these would constitute our indicators.

It was possible for studying and classifying problems to come too early. The study benefited as in the preceding design from the good mathematical knowledge developed in the first 11 modules. But the questions of proportionality had become more difficult to collect and master because of local singularities that appeared there. We observed that augmenting the collection of classical problems presented in different environments resulted in an increase in the volume of the vocabulary and metaphors brought into play and an improvement in the execution of algorithms, but also in a diminution of the students' capacity to verify and explain their calculations.

The phenomenon seemed all the more marked in that we had taken a lot of care to get students to use manipulations, formulations and explanations that were more precise and better based on their actions. Classical knowledge about fractions passes for "concrete". In fact, it consists of metaphors, verbal connections and cultural habits often stripped of real concrete meaning, unlike the knowledge that we developed. The complexity of the concept comes from its roots in the culture, the explosion of the collection of meanings and the absence of a sufficient use of unifying instruments.

Later on the teachers did not always maintain the insertion of the classification of problems before the last two modules, but it did make the last two appear to be a clarification necessary for the teachers and for the students.

Owing to a shortage of researchers and observers, the complementary sessions on the study of problems could not be undertaken with the normal and necessary scientific environment of the COREM. The effects of these modules, when they were carried out, therefore could not be collected and analyzed.

Modules 14 and 15 return to and complete the mathematical study suspended since the end of Module 10. The students unify the conceptions of multiplication and division of fractions and rational numbers that they have learned, by putting them back into the multiplicative group of the rational numbers.

Module 14: Composition of Linear Mappings

Lesson 1: The Pantograph

Materials[9]

One pantograph for every pair of students (the pantographs have different scales)
At least 4 sheets of unlined paper per pair of students
Tape to stick the paper to the desk
One eraser per pair of children

The pantographs are distributed before the lesson begins (Fig. 2.9).

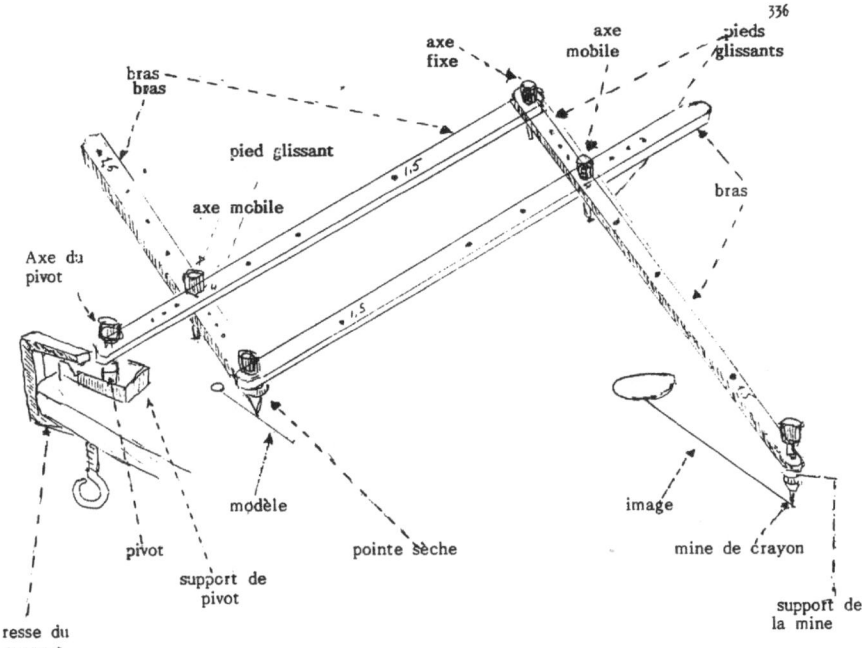

Fig. 2.9 The pantograph

---

[9]This activity can take place in the context of a Social Studies class – a drawing, for example.

Introduction of the Pantographs

(a) Instructions: "These gadgets are pantographs. Some of you may have seen them or even used them. They let you reproduce designs. So you are going to make a design on one of your sheets of paper and reproduce it on the other one."
(b) Development: The children work in pairs. There is always a moment of hesitation (as is often the case when children are confronted with a new situation or an instrument that they have never used.) But very soon they organize themselves, make a design and figure out how to use the pantograph: the pivot mustn't move, nor the paper (which is taped down), the pointer follows the model, the pencil draws the image.

As soon as they have produced a few designs they ask to modify the form of the pantographs, and the teacher says they may. They modify the scale, start drawing again, are surprised by some of the modifications and amused by others.

This free manipulation of the pantographs can hold their full attention for 30 or 45 min.

Discussion of observations

After this playful phase, the teacher gathers them to make a collection of observations based on the designs and their reproductions.

(a) Instructions: "What did you notice?"
(b) Process: The children make remarks and hypotheses:

- You can enlarge or shrink by exchanging the pencil and the pointer.
- The shape of the image doesn't change no matter how you set up the pantograph.
- The "enlargement" or "shrinkage" varies with the scale of the pantograph.
- The numbers beside the holes indicate the amount of enlargement or shrinkage.

This remark is immediately verified by the children who made it: they display their model and its reproduction, measure a segment on the model and the corresponding segment on the reproduction and write the measurements on the board.

	Design (cm)	Reproduction (cm)
Example:	3,2 $\longrightarrow$	5,4

And calculate 3.2×1.5 (if 1.5 is the number by the hole on the pantograph)
$3.2 \times 1.5 = 5.25$!
General astonishment! "They made a mistake!" The teacher suggests that all the students check the operation – notebooks, calculations … it really is 5.25!
So there is a 1.5 mm error!
The children make comments: "It's not a big error!" "It's bound to happen, because the pantograph isn't very precise." Many of them then want to check whether their own reproduction was better realized, and by pairs they verify using the above procedure. They get quite competitive: each one hopes to have succeeded better than the others!

If the pantograph isn't set up as a parallelogram, the image is deformed.

It always happens that one group sets up its pantograph without paying attention to the numbers by the holes. When the time comes to share their results they are a little embarrassed to present their designs because sometimes they haven't figured out what caused the deformation. The class often gets a good laugh out of the reproductions, which can be bizarre shapes.

Others who figured it out in the phase of free manipulation come to the aid of their comrades by saying that the same thing happened to them, but they noticed that the scales were badly set up and fixed them.

Results All the children know how to use the pantograph. They also all understand the remarks that were made and know thereafter how to get what they want out of a pantograph.

Lesson 2: Composition of Mappings: First Session

Materials

Two pantographs per group of two or three students, one set to enlarge by a factor of 3, the other by a factor of 1.5

Three sheets of paper of different colors per group and for the teacher.

Presentation of the Situation

"On the back of this white piece of paper I made a design. Then, with this pantograph I reproduced the design on the blue sheet of paper. Then finally I reproduced the design on the blue paper on this yellow paper using this pantograph."[10] (The designs are on the backs of the pages, so the students don't see them.)

1. Qualitative predictions

   "What can you guess about these designs? What can you say without seeing them?"

   The teacher can count on the following answers: "They will look like each other" "They will be enlarged or shrunk", "You have to see how the pantograph is set up."

   So the teacher demonstrates where the pointer and the pencil are on the two pantographs. The children then say "The designs are enlarged. The yellow one is the biggest."

---

[10]The teacher's pantographs are also set to scale factors of 3 and 1.5.

2. Quantitative predictions

"In a moment you are going to do the same thing: you will make a design on the white paper, reproduce it on the blue paper with the first pantograph, then reproduce the one on blue on the yellow paper using the other pantograph.

But first, I am going to give you two dimensions of my model:

4 ———————————————→

2.5 ——————————————→

(The teacher writes these measurements on the board.)

Can you predict the corresponding dimensions on the yellow paper?"

The students say that they need more information and request either by how much the pantographs enlarge or a corresponding dimension on the yellow paper.

Presentation of a Game: First Try

(a) *Assignment*: "You are going to play a game: in your notebook you are to write the information that you want. I will give it to you. After that it is a matter of making predictions: you choose some numbers that designate measurements on the model and you predict the lengths of the corresponding segments on the yellow sheet.

You must write these numbers in a table.

Example:

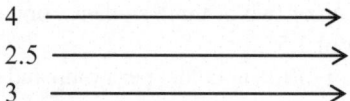

4 ———————————————→

2.5 ——————————————→

3 ———————————————→

When you have predicted the corresponding measurements, you can verify if your prediction is right by using the pantographs.

If you choose a whole number and the prediction is right, you get one point.
If you choose a decimal number and the prediction is right, you get three points.

(b) *Development:*

The children work in groups of two or three. The teacher suggests, if they haven't thought of it, that each one calculate a measurement (because the points can be added up). That way they can have more because in the groups the children have a tendency to calculate the same measurement together, which slows the calculations and at the same time limits their number.

While the students are making their predictions, the teacher prepares the following table.

	Predictions		Correct predictions	Points
	Whole	decimal		
Group 1				
Group 2				
Group 3				
Group 4				
Group 5				

(c) Verification, done in two parts:

1. First collectively for 4 and 2.5: the teacher has one child from each group come to the board.

    One of the children draws a 4 cm line on a white paper; another uses the pantograph to make the first image on blue paper; another does the second image on yellow paper. Finally still another measures the images and gives the results to the teacher, who puts them on an enlarged representation on the board:

4 ⟶ 12 ⟶ 18

    The teachers asks, "Who got these measurements?" Often there are errors caused by the pantograph, which gives rise to discussions. A consensus is established: predictions that are within three tenths will be accepted: for example, if they predicted 18 and came out with 17.7 or 18.2 their prediction would count as correct.

2. For the other predictions, verification is done in each group with pantographs. A child from a concurrent group comes to check. If the activity takes too long (because the children are still not very adroit at using the pantographs) the teacher can switch to a simple collective verification like the ones for 4 and 2.5.

    The teacher collects the results in the prepared table and scores the points.

Game: Second Try

(a) *Instructions*: "You saw that it was possible to predict the measurements of the last image with calculations, some slow, some fast. You are going to try a second time. You will discuss with your group how to find the speediest way to calculate the results, so that you can make the most predictions possible. Here is a list of numbers. You are to choose the numbers on this list that you want – as many as possible of them. You can even add some if you are very swift and if you want to."

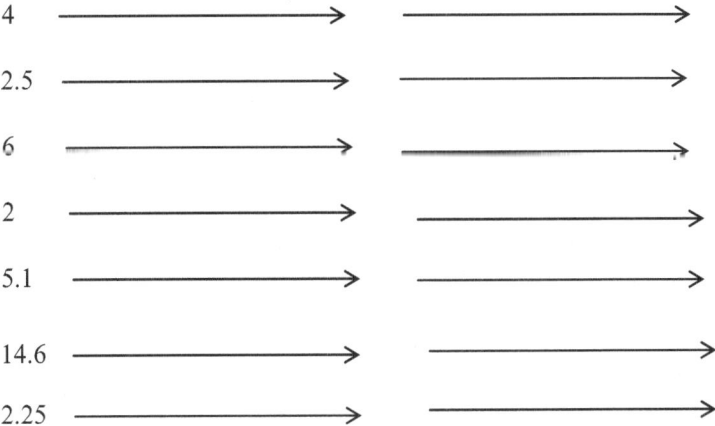

4

2.5

6

2

5.1

14.6

2.25

(b) Development: The children continue to work in groups, dividing up the work.
    Collective synthesis: correction of results, inventory of methods
(a) Correction of results. The teacher takes the results the children have found and
    puts them on the board, correcting them in the process, then gives out the points
    in a way that lets the children know which team won. After that the class makes
    an inventory of the methods they used:
(b) Inventory of methods: One child from each team comes to the board to explain
    his method. Clearly not all will come, because after each demonstration, the
    teacher asks who else used the same method.
    First method observed: Calculation of intermediate values by linearity

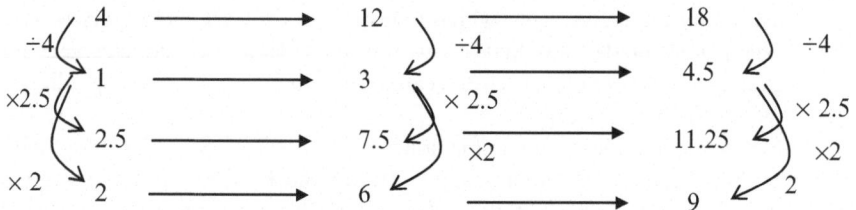

Second method: Calculation of intermediate values as a product:

$$\begin{array}{ccc} \times 3 & \times 1.5 \\ 4 \overset{\frown}{\qquad} 12 \overset{\frown}{\qquad} 18 \end{array}$$

$$\begin{array}{ccc} \times 3 & \times 1.5 \\ 2.5 \overset{\frown}{\qquad} 7.5 \overset{\frown}{\qquad} 11.25 \end{array}$$

Third method: No intermediate calculations:

$$\begin{array}{cc} \times 4.5 \\ 4 \overset{\frown}{\qquad\qquad} 18 \end{array}$$

$$\begin{array}{cc} \times 4.5 \\ 2.5 \overset{\frown}{\qquad\qquad} 11.25 \end{array}$$

Obviously, the children who used the third method were able to calculate all
the results quickly, while those who used the two others, especially the first,
didn't often get to the end of the list. They recognize that the last method is the
fastest, but there are always a few who ask: "Why did you multiply by 4.5?", to
which some answer "Because $3+1.5=4.5$" and others say "What we did was to
multiply 3 by 1.5". The latter is generally not accepted, however, because the
children can see immediately without calculating anything that $3+1.5=4.5$,
whereas to multiply 3×1.5 they have to carry out a calculation (even if it is a
mental one.) The problem therefore stays open.
The teacher writes on the board the following conclusion:
$(\times 3)\ F\ (\times 1.5)=(\times 4.5)$, where the F stands for "Followed by"
(c) Open problem: "Can you predict what enlargement you'll get from two panto-
    graphs set to 3.5 and to 2? Think about it and give your answer next time."

A very anarchical discussion takes off among the children: some think it is 5.5, others disagree. The teacher doesn't take part, and tells them to think about it for the next day and above all to find a proof that what they are saying is true. The session ends in a state of suspense that excites the interest of the children and sets them up for the next activity.

Results The children have composed two linear mappings. They have anticipated the result, found several methods, and chosen the shortest. They have discovered that they can cut down on calculations by replacing two linear mappings by some linear mapping, but they don't know yet how to calculate it.

Lesson 3: Composition of Linear Mappings: Designation of Composed Mappings

Search for a solution to the open problem and validation

1. Review of preceding activity by the teacher.
   "Last time we enlarged a model with the pantograph set to 3. Then we enlarged the first image with a pantograph set to 1.5. We saw that the enlargement that would let us go straight from the original to the second image was (×4.5) and we wrote that

$$(\times 3)\ F\ (\times 1.5) = (\times 4.5)$$

2. Open problem
   (a) *Instructions*: "At the end of the last class, I gave you the following assignment: I used the pantograph to make the enlargement (× 3.5) F (× 2) and I asked you what linear mapping could replace these two mappings. If you found a solution, write it in your notebook."
   (b) *Development*: The children write an answer. After two minutes, the teacher asks what answers they wrote in their notebooks, and writes them on the board. There are always two:

$$(\times 3.5)\ F\ (\times 2) = (\times 5.5)$$

$$(\times 3.5)\ F\ (\times 2) = (\times 7)$$

   So they proceed to a collective verification.
   (c) *Verification*: "How can we know which one is the right answer?"
      The children propose to verify the mappings with whole numbers or decimals.
      Verification on a whole number measurement, for example 8:

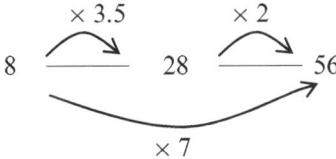

The mapping that gives 56 as the image of 8 is indeed × 7. The teacher then has the students calculate what the image of 8 would be under the mapping × 5.5. That one gives 44, which doesn't correspond to what they got using intermediate steps. The students thus see clearly that

$$(\times\,3.5)\,\mathrm{F}\,(\times\,2)=(\times\,7)$$

Verification on a decimal number, for example 1.5

The students first calculate the values using the two steps × 3.5 and × 2. Then the teacher has them calculate 1.5×7 and 1.5×5.5, and again the former corresponds to the image found and latter doesn't. This solidifies the conclusion that

$$(\times\,3.5)\,\mathrm{F}\,(\times\,2)=(\times\,7).$$

(d) Rule of composition for two mappings

The teacher has them formulate the rule that they found after these two verifications:

"To find the linear mapping that replaces two linear mappings, you have to multiply the mappings."

Verification on a very simple measurement: 1

The teacher asks the children if they couldn't verify the same thing but avoid messy calculations where they might make mistakes.

"What's the really simple measurement that you could start with and verify that the mapping you found is right?"

A few children suggest 1 (if nobody thinks of it, the teacher suggests it) and they try it right away. One of the students who proposed it comes to the board and writes:

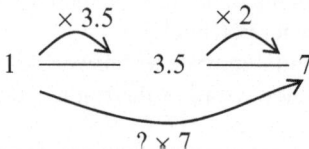

Verification of the rule for any sequence of enlargements or reductions

a) *Instructions*: "Now you know how to find a mapping that lets you replace two successive mappings. But does that work if you have more than two mappings? To know that, you are going to calculate a lot of examples that I am going to write on the board and after that you can say whether the rule is general."

Examples:

$(\times\,1.75)\,\mathrm{F}\,(\times\,1)\,\mathrm{F}\,(\times\,0.5)$
$(\times\,3)\,\mathrm{F}\,(\times\,2.5)\,\mathrm{F}\,(\times\,1.75)$
$(\times\,0.125)\,\mathrm{F}\,(\times\,5)\,\mathrm{F}\,(\times\,1.5)\,\mathrm{F}\,(\times\,2)$
$(\times\,4.5)\,\mathrm{F}\,(\times\,0.2)\,\mathrm{F}\,(\times\,0.7)$
$(\times\,2.7)\,\mathrm{F}\,(\times\,4.52)\,\mathrm{F}\,(\times\,0)\,\mathrm{F}\,(\times\,0.425)$

Development: The children do the calculations in groups of two or three. The teacher assigns one or two series of mappings to each group, who are to replace the series by a single mapping, proving that it is correct.

c) *Collective correction*:

One student from each group goes to the board to show she has done, under the critical eye of the others, who follow the calculations attentively. They choose 1 as the measurement to start with.

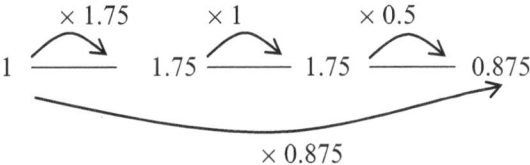

First example: ($\times$ 1.75) F ($\times$ 1) F ($\times$ 0.5)$=0.875$

The rest of the examples are done the same way. The teacher takes the opportunity offered by this correction process, which gives the students no trouble at all, to get the students to discover the properties of these compositions of mappings: commutativity, associativity, the role of 1, the role of 0, etc. (Many of the children, in fact, start in on long, complicated calculations before noticing that there is a ($\times$ 0) in the course of the mappings.)

The teacher points out that no matter what the mappings are (enlargements or reductions), and no matter how many of them there are, they can always be replaced by a single mapping by multiplying them all together.

She points out to the children that these multiplications are different from the ones they already knew (cf. Module 8, activities 2-4 – finding the image of a measurement under a decimal number mapping.)

Individual exercises:

$$13.4 \ (\times 3.5) \ F \ (\times 1.5) \ F \ (\times 4) \ \xrightarrow{?}$$

$$25.86 \ (\times 3.5) \ F \ (\times 1.5) \ F \ (\times 4) \ \xrightarrow{?}$$

$$11 \ (\times 3.5) \ F \ (\times 1.5) \ F \ (\times 4) \ \xrightarrow{?}$$

In the course of correcting these the teacher takes note of the methods. There are still children who carry out the intermediate calculations, thus making mistakes and proceeding much less swiftly than those who go directly to the mapping ($\times$ 3.5 $\times$ 1.5 $\times$ 4). This then provides an occasion for the children to become conscious of the utility of replacing several linear mappings with a single one and of making use of the rules that they learned in the course of the activity.

Results This activity presents no difficulties at all. It not only gives the children a chance to multiply some decimals and rediscover the meaning of this multiplication, but permits them, thanks to the rules discovered, to save some calculations and to design new mappings.

Lesson 4: Different Ways of Writing the Same Mapping

Materials
One pantograph
Review of the rule of composition for mappings

(a) *Instructions*: Here is a sequence of linear mappings:

$$(\times 1.5) \text{ F } (\times 2) \text{ F } (\times 2.5) \text{ F } (\times 3) \text{ F } (\times 4)$$

Can our pantograph make that enlargement? If so, what linear mapping could one substitute for the sequence of enlargements?

(b) *Development*: The children work on their own in their scratch notebooks, working as fast as possible. The teacher invites them to find the fastest possible calculations.

(c) *Correction*: After 3 min, there is a collective correction. For that, the teacher sends one student to the board to write

$$(\times 1.5) \text{ F } (\times 2) \text{ F } (\times 2.5) \text{ F } (\times 3) \text{ F } (\times 4) = 1.5 \times 2 \times 2.5 \times 3 \times 4.$$

The child explains how he did it and very often the others propose various solutions from their places:

"What I did was to do everything in my head that I could:

2 times 1.5 makes 3
3 times 3 makes 9
9 times 4 makes 36

and all that's left is to multiply 36 by 2.5, and you can do that in your head, too:

2 times 36 makes 72; half of 36 is 18, 72 plus 18 makes 90!"

Another one says "I started off with 3 times 4, that makes 12, then 2 times 12, that makes 24. 24 multiplied by 1.5 makes 24 + 12 = 36, and all that's left is 36 times 2.5"

This goes on until all the procedures have been stated. This development lets all the students discover all the possible methods of doing the calculation and find the fastest.

At the end, someone writes on the board

$$(\times 1.5) \text{ F } (\times 2) \text{ F } (\times 2.5) \text{ F } (\times 3) \text{ F } (\times 4) = (\times 90).$$

Different ways to write the same mapping

(a) Presentation of the problem and instructions:
   "What can the pantographs that you have been using do? (They can either enlarge a model or shrink a model). You have already used these different possibilities. So you should be able to give the meaning of:

   'Shrink by 3'.
   Would you know how to do that with this pantograph and write in your notebook what this mapping does?"

(b) Development
   The teacher gives the children a moment to think and then asks two of them to come an carry out with the pantograph, in front of all the class, a reduction by 3.
   They set the pantograph to 3 and put the pencil between the point and the pivot, helped if necessary by remarks from other children. Then the teacher has them draw a 9 cm line segment on a piece of paper which she then tapes to the board and with the help of the pantograph they have set up, they draw the image of the segment. The class makes comments out loud:

   "The image is 3 cm long because it should be 3 times smaller than the model. Maybe it's not quite exact on the design…"

   The teacher reminds them of the last question he posed: "Would you know how to write in your notebook what this mapping does?" and lets the children think about it a minute or two. Then one of them comes to the board and writes:

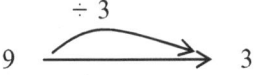

   The teachers adds some measures and asks the students to complete the following table:

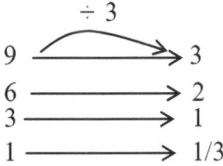

   He proceeds to a rapid collective correction and detaches the last pair on the board:

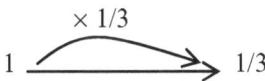

"How could we designate the mapping that takes 1 to 1/3?"
The children spontaneously answer "(× 1/3)"

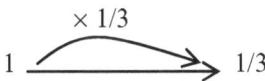

But they have to check that it really is the same mapping, if it works also with the preceding measurements: 6, 3 and 9, which a child promptly does:

$$6 \xrightarrow{\times 1/3} 6/3 = 2$$

$$9 \xrightarrow{\times 1/3} 9/3 = 3$$

It gives the same images as the mapping ($\div$ 3). So we can write ($\div$ 3) = (x 1/3)

Conclusion

The teacher says: "This mapping is called "dividing by 3" or "multiplying by 1/3". It can be written as "divide by 3" or as "multiply by 1/3". There are lots of ways to write it.

Other names for the mappings ÷4 and ÷2

(a) Instructions: "We are going to try to write some other mappings in a bunch of ways. What reductions can we make with our pantograph using only whole numbers?

   –   You can shrink by 4 or by 2

      "How can we write what those mappings do? Who knows how to find several ways to write what they do?"

(b) Development: The children work on it a moment in their scratch notebooks. Each one makes it a point of honor to find a different name. The teacher proceeds quickly to a collective correction so as to keep the interest lively.

   Children take turns coming to the board to write:

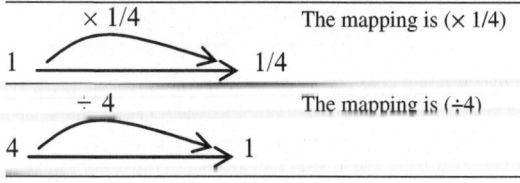

"Could we find another one by replacing the fraction 1/4 by a decimal number?"
The children calculate quickly in two different ways:
First way (most often used): 1/4 = 25/100 = 0.25
Second way: by division 1 ÷ 4 = 0.25

(c) Validation of the names

   Instructions: "We just found three different names. ($\div$ 4), (x 1/4) and (x 0.25) Now we need to check whether they really are the same mapping. So we are going to apply them to some other numbers: 2.5, for instance."

Development The teacher suggests that the class divide up the tasks to save time: one row calculates with (÷ 4), one with (× 1/4) and the third with (× 0.25).

Correction: One child from each row comes to the board and writes the calculation for their row's mapping of 2.5. All find 6.25.

So the mappings really are all the same, because they give the same image. So we can write:

$$(÷ 4) = (× 1/4) = (× 0.25)$$

This equation is written up and left on a corner of the board or on another board.

The teacher goes through all the same steps for reduction by 2, getting (÷ 2), (× 1/2), and (×0.5), and having them check by applying all three to 7.8.

He writes beneath the previous equation:

$$(÷ 2) = (× 1/2) = (× 0.5)$$

Other examples: generalization

(a) *Instructions*: "If you had a pantograph that shrank things by 5, or 6, or 9 do you think you could find other names for those mappings?"
(b) *Development*: The children work in their scratch notebooks. Some of them, by analogy, immediately write

$$(÷ 5) = (× 1/5) = (× 0.2)$$

Others still need to calculate the long way:

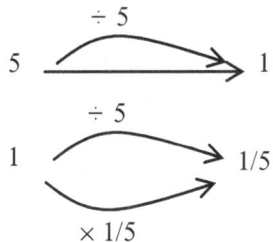

When it comes to (÷ 6) = (× 1/6) = ?, they hesitate because they observe that "It doesn't come out right!" The teacher decides with them that they will write it as

$$(× 1/6) = (× 0.1\underline{6} ...)$$

Use of the different names

To finish up, the teacher organizes a brief session of mental calculation. He writes on the board the following operation:

$$4 × 0.25 =$$

and gives them 20 seconds to figure it out without writing it in vertical format.

The students hesitate, start the multiplication mentally and protest when the teacher stops them after 20 seconds. Only one or two have found the answer.

The teacher points out the equalities that are still written on the board: nobody had thought of replacing (x 0.25) by (÷ 4)

$$4 \div 4 = 1.$$

The students catch on and ask for some more calculations to do. The activity finishes up as a real game.

$$18 \times 0.5 = ?$$
$$25 \times 0.2 = ?$$

Results The students know several ways to write the same mapping.

Lesson 5: Rational Linear Mappings

Presentation of the Problem

(a) *Instructions*: "What are the whole number linear mappings you can do with our pantograph?"

The teacher writes on the board, as directed by the students,

$$(\times 2) (\times 3) (\times 4)$$
$$(\div 2) (\div 3) (\div 4)$$

"I'm going to take the pantograph that enlarges by 3 and draw an image with it. Then with the pantograph that shrinks by 2 I will shrink the image I got and make a second image. Do you know how to write what I did?"

The class answers: "You multiplied by 3 and then divided by 2." And one of them is invited to write these two successive mappings on the board:

(b) *Problem posed.* "Can we combine these two linear mappings to get a fractional or decimal mapping?"

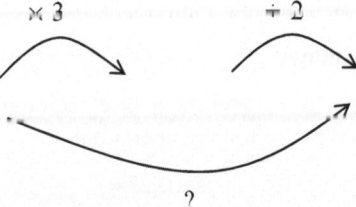

(c) *Development:* This is done collectively with the teacher. The children suggest taking a number and naturally choose 1 (because that is the number that has always had priority.)

The Teacher Writes

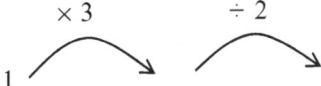

and asks a child to come to the board and complete the diagram, putting in the intermediate numbers:

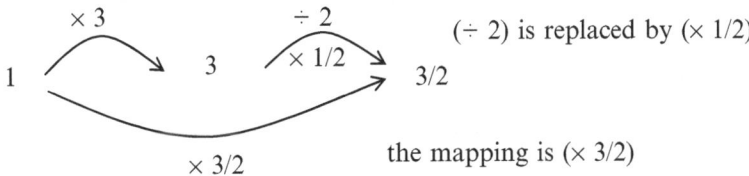

($\div$ 2) is replaced by ($\times$ 1/2)

the mapping is ($\times$ 3/2)

The teacher writes up the conclusion: ($\times$ 3) F ($\div$ 2)=($\times$ 3/2)=($\times$ 1.5), the last having been rapidly calculated by the children.

It is thus possible to replace two whole number linear mappings with a fractional or decimal linear mapping.

A search for all the rational linear mappings the pantograph can produce

(a) Instructions: "Use the same method to find everything else that we could do with our pantograph by combining all the whole number mappings."
(b) Development: The students work a little while in their scratch notebooks. After 5 min, the teacher asks them to come write on the board what they have found. This gives a sequence of mappings:

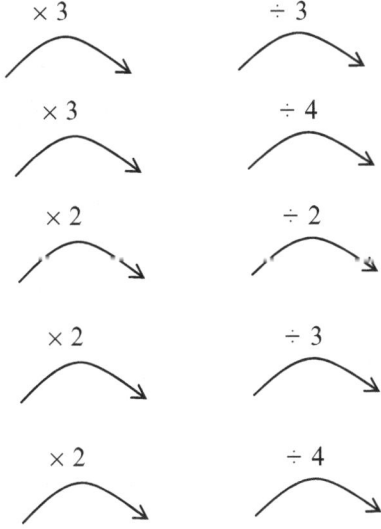

etc.

Making a Table

(a) *Instructions*: The teacher suggests making a double entry table to avoid missing anything or duplicating anything.

	× 2	× 3	× 4	÷ 2	÷ 3	÷4
× 2						
× 3						
× 4						
÷ 2						
÷ 3	x 2/3					
÷ 4						

He tells the children to complete the table, checking their results each time.

Example: (× 2) F (÷ 3)

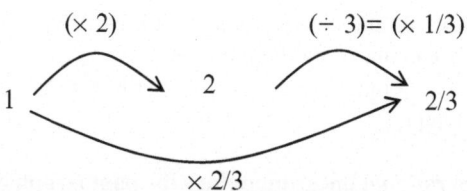

(b) *Development*: The children work individually. They complete the table they have drawn in their scratch notebooks and don't write in a result until they have checked it as before.

Remarks: While they are filling in their tables they often make comments out loud:

"You don't always get a fraction!"
"You get some things we already learned!"

    (They are talking about (× 4) F (÷ 2), for instance, which they saw at the beginning of the year in an activity on functions corresponding to operations on natural numbers.)

(c) *Correction*: After 5–8 min of individual work, the teacher organizes a collective correction. The children take turns coming to the board to fill in the table that the teacher has drawn for them.

    In the course of doing it they make more comments like

"If you do (× 3) F (÷ 3) or (× 4) F (÷ 4) it's as if you did (× 1) or (÷ 1)"
"If you do (× 3) F (÷ 2) = (× 3/2) it's the same thing as if you did (÷ 2) F (× 3) = (3/2)"

    This way they discover the commutativity of mappings. They verify that it is true for all cases, using the usual format.

After checking all of the entries, the students formulate the rule for composition of mappings:

"Any decimal or fractional linear mapping can be gotten by doing two whole number mappings in a row. The number that is multiplied is always on top, and the number that is divided is on the bottom."

Application exercises, done individually in mathematics notebooks

Instructions:

1. Find the rational and decimal linear mappings when it can be done, to replace two whole number mappings:

   Example: $(\times 7)$ F $(\div 2) = (\times 7/2) = (\times 3.5)$
   $(\div 5)$ F $(\times 4) =$
   $(\times 8)$ F $(\div 5) =$
   $(\times 4)$ F $(\div 5) =$
   $(\times 12)$ F $(\div 12) =$
   $(\div 5)$ F $(\times 5) =$

2. Find the mapping that is missing in each of these:

   $(\times 5)$ F $(\ ) = (\times 5/3)$
   $(\div 4)$ F $(\ ) = (\times 7/4)$
   $(\ )$ F $(\times 2) = (\times 2/9)$
   $(\ )$ F $(\times 3) = (\times 1)$
   $(\div 5)$ F $(\ ) = (\times 1)$

Results This activity gives the students no trouble at all. They all understand and know how to do the individual exercises. There will, however, be some errors to correct. The activity finishes with a collective correction (which can be done at the beginning of the next session if time runs out.)

Module 15: Decomposition of Rational Mappings. Identification of Rational Numbers and Rational Linear Mappings

Lesson 1: Decomposition of Rational Mappings

In this activity the teacher asks questions that have not been directly addressed, but for which the students can almost instantly find an answer (Socratic *maieutique*)

Decomposition of a Rational Mapping into Natural Number Mappings

Teacher: "I want to carry out a 3/4 enlargement (which is a diminution) but I don't have a pantograph that does × 3/4. With the pantographs we have would we be able to do the enlargement?"

The teacher takes care not to have his request confused with an "enlargement by 3/4", in the sense of adding 3/4, which would in fact be × 7/4.

When the students propose their answers orally the teacher writes them on the board using arrows, as was habitually done since the reproductions of the Optimist (Module 9).

For example:

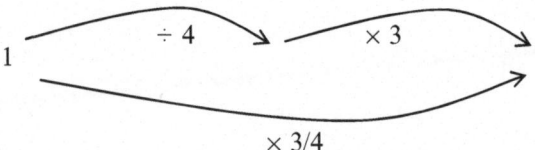

$$\times 3/4$$

"Without arrows we could write $(\div 4)F(\times 3)$", which she reads "divide by 4 and then multiply by 3"

She continues with similar questions: "× 7/12", "take 3/5 of something", then "× 3.67" The children hesitate a moment and them some shout out (instead of writing) "You have to change the 3.67 into a fraction!"

$$3.67 = 367/100 \rightarrow (\times 367).(\div 100)$$

Decomposition of the Reciprocal

"To get from a model to its image I used the pantograph × 4/7. How could I do the reciprocal mapping with whole number pantographs?"

The solution is obvious materially, because pantographs are invertible. The students *explain the action* and invert it by replacing the multiplications by divisions. It's not enough to reverse the arrows!

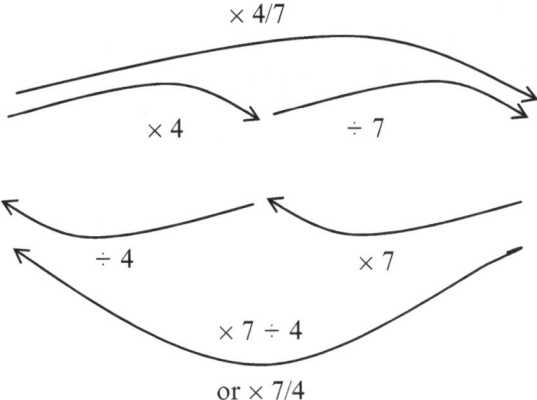

or × 7/4

The students rediscover a result they already knew in another context. The teacher reminds them of this method of representing measures and operations on those measures:

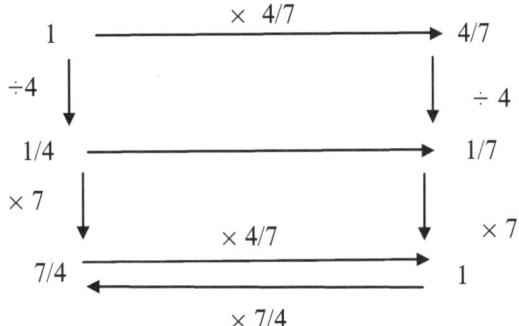

"You can find the reciprocal of a fractional linear mapping by decomposing it, taking the reciprocals and recomposing it."

The teacher then studies in the same way the reciprocal of a ratio:

"Kafor coffee is a mixture: 4/7 of its weight is made up of Arabica coffee. What operation would let us figure out how many pounds of Kafor coffee we could make with various different weights of Arabica?"

Decomposition of the mapping x1; inverse mappings

3 ⟶ 3

2.5 ⟶ 2.5

4/7 ⟶ 4/7

"Here is a mapping. What is it?"
"It's the mapping (×1)."

"Can we decompose it?"

First reaction from the student: "Can't do it!"

The teachers pushes them: "There isn't any pair of mappings that can be replaced by ×1?"

Students: "Oh, yeah! You can enlarge by 3 and shrink by 3…"

The teacher illustrates their suggestion with

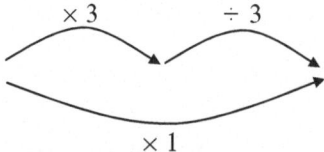

After a few more illustration she poses the question: "There is a report that the number of accidents has increased 5 % over last year. By how much they decrease this year in order for next year's number to be the same as last years?"

Some of the students first try to use the arrows directly:

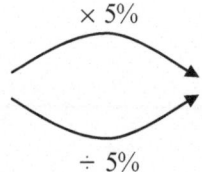

But when they carry out the calculations they discover that decreasing by 5 % doesn't work. So they pose $1 + 5/100 = 1 + 0.05 = 1.05$ and represent the calculations to carry out as

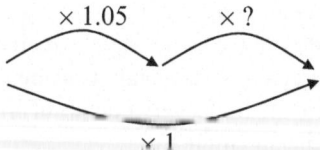

Results The students become familiar with new vocabulary and situations. For many of them, handling compositions of mappings seems easy but risks becoming formal, and too quickly escaping from the control that the students need to exercise by verifying the meaning.

Thus for the teacher this is not the moment to institutionalize these new ways of calculating and still less to require mechanical reproduction of them.

Important Remark, 2008

This warning to the teachers was essential and needs to be explained to today's reader to avoid misunderstanding the nature of the teaching and of the practices described here. Up until this moment, the arrows have been used exclusively to designate mathematical objects: either correspondences (not necessarily numerical ones), or natural relationships between numbers in the same set of measures (for example natural differences or later natural ratios, but never both at the same time). Later they were used to indicate rational linear mappings (horizontal arrows for the enlargement x 1.75, etc.) Now they are used in showing that natural ratios like (x3) or (÷3) can be replaced respectively by (÷1/3) or (x1/3). But they have remained a free means of expression and have not been the object of any teaching or any evaluation. In Module 15 the study of compositions of mappings gives them a new status. They become an instrument of analysis, of calculation and even of proofs, and thus also an object of study and discussion. Their disposition may change – but they do not have the properties of a good model. It is therefore essential that the teacher not treat them as a piece of mathematical knowledge, that he not teach them lest he launch a metadidactical slippage that would be difficult to control. They should be used only as a means of expression, a prop for reasoning whose validity the student checks by reference to the actual meaning. It should be well noted that formal operations, their representations by arrows and the reasoning presented in these chapters **are not pieces of formal knowledge to be taught in the classical sense.**

Lesson 2: The Meaning of "Division by a Fraction"
(Summary of Lessons)

"Would you know how to give a meaning to the operation 4 ÷ 3/5?", asks the teacher.

What the students need to do is

First interpret the formula by a real situation like the ones they have encountered in solving problems. For example:

- You divide a 4 m long ribbon into parts that are 3/5 of a meter long
- You buy 3/5 of a meter of ribbon for 4 francs. What is the price of a meter of ribbon?
- A rug with area 4 m² has width 3/4 m. How long is it?

Then with the help of their schemas on quantities, relations and mappings they figure out the operations to carry out.

The teacher collects the problem statements they are trying to invent and helps them pull their ideas together, then organizes a discussion among the students about the statements they have come up with. A statement may be interpreted in a variety of ways. The teacher tries to get them to reformulate each of the possible interpretations of the givens as measure, then as linear mapping.

This type of activity is organized by a didactical schema known as a "tournament of problem statements", which is described in Module 12. A few examples of similar questions lead the students to be interested in the interpretation of the value being sought as a ratio of amounts or as a linear mapping.

In this lesson the products of fractions are finally conceived as products, that is, compositions of direct or inverse linear mappings that make it possible to ask questions like: What linear mapping does 7/9 ÷ 3/5 represent?

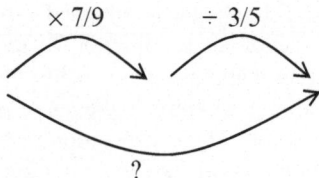

The decomposition of (× 7/9) into (× 7) followed by (÷9) is one they know well

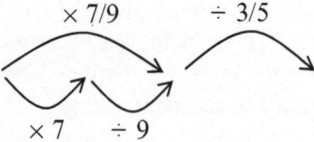

Then (÷ 3/5) is the reciprocal of (× 3/5)

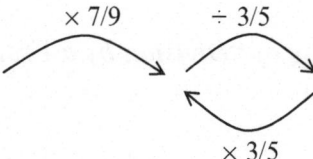

And that in turn can be decomposed into (× 3) followed by (÷5)

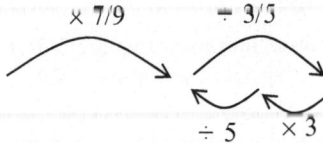

Their reciprocals can also be calculated:

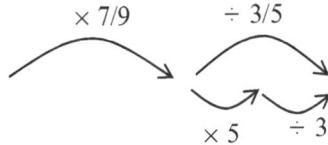

Thus the sequence of calculations is established as $(\times 7)(\div 9)(\times 5)(\div 3)$, and since the order of linear mappings can be modified, we get $\times 7 \times 5 \div 9 \div 3 = 35 \div 27$

Commentary

This "proof" calls for some details about the development of the lesson:

1. No, it was not a response given individually by each child in response to a test question or an individual exercise.
2. The teacher offered the exercise for students to reflect on individually for a while, then collected the students' suggestions. Everyone achieved the first step, and the second was routine.
3. The third step gives no difficulties to certain students, but although they respond with "÷ 3 followed by x 5" it is for bad reasons that they can't justify: they broke up the fraction as they would have done with x 3/5, inverting the last sign. The teacher says nothing, but other students express concern. The class is not in the habit of calculating without knowing what they are doing.
4. The teacher guides the discussion: "You don't know how to decompose ÷3/5?" Some of the students recall recent calculations about this kind of linear mapping by taking numbers.
5. The teacher suggests that they know how to decompose the reciprocal. The students then develop the method. Each step is a sort of rapid individual exercise.

So here is a matter of a sequence of "exercises". This would be a problem for students who wanted to solve it by themselves. Certain of them, stimulated by challenges from others, could get there by themselves, but at what price and for what profit (for themselves or the others)? The solutions of the steps are exercises that the teacher rapidly proposes and checks.

This problem is not a lesson and its solution is not a piece of knowledge to be learned. It is simply an occasion for using the knowledge that is in process of being learned and making it more familiar and more easily available. Not every student solves every exercise, but they will see a certain number of them again.

Lesson 3: Division of Decimals

A Mapping For the Calculation of Decimal Numbers

(a) *Assignment*: "Would you know now how to find a meaning for this division: $1.38 \div 4.15$?" (the operation is written on the board.)

(b) *Development*: This phase proceeds like the preceding one: First, time for the students to reflect and try things out in their scratch-notebooks. Then an alternation of individual and collective reflections in the course of which the teacher has them explain the meaning of this division: "We have to find the image of 1.38 under the mapping $\div 4.15$", and has someone write on the board (or writes herself):

$$1.38 \xrightarrow{\ \div 4.15\ } ?$$

The children, who by this time are well trained on this kind of exercise, suggest writing 4.15 as a fraction: 415/100

First step:

$$1.38 \xrightarrow{\ \div 415/100\ } ?$$

Referring to the activity of 15.2.1, the teacher asks: "What mapping can we replace $\div 415/100$ by?", and writes the mapping (or has a child write)

$$\div 415/100 = \times 100/415$$

Second step:

$$1.38 \xrightarrow{\ \times 100/415\ } ?$$

Third step:

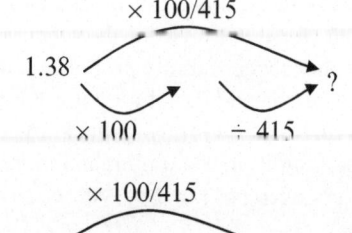

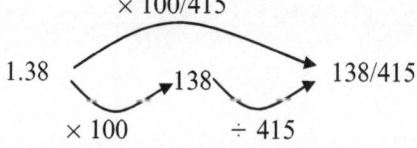

The children calculate 138/415 in their scratch notebooks

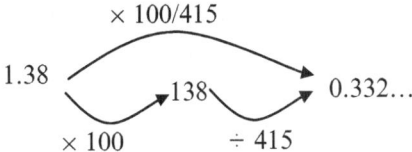

Conclusions and Installation of the Algorithm

The teacher asks what calculations they had to make to find $1.38 \div 2.14$. "We had to multiply by 100 first (138) and then divide by 415."

The teacher calls the students' attention to this new method of "division" with a different meaning from the one they knew before (see modules 12 and 13), which they can now calculate rapidly no matter what the numbers (without writing the successive subtractions or finding the intervals by trial and error.)

With a few remarks connecting what they have just learned with the notation and format they used previously, this concludes the curriculum.

Chapter 3
The Adventure as Experienced by the Teachers

In Chap. 2 we presented an adventure in the learning of fractions and decimal numbers, with our perspective being that of the students who were doing the learning. In this chapter we step back and have another look at the same adventure. We will first set out the context: when and why the curriculum was created, the research questions underlying it and the school and research environment in which it was embedded. With that in hand, we will look again at the adventure itself, this time from the perspective of the teachers. In Chap. 4, we will take one further step back and examine the mathematical context and the reasoning behind the mathematical choices made in constructing the curriculum from the perspective of the researchers.

Background of the Project

Two elements of the background that were described in Chap. 1 are sufficiently pertinent to this chapter that we will start by reproducing them: The lessons described in Chap. 2 took place at the COREM (Center for Observation and Research on Mathematics Teaching), which was a regular public school in a blue collar district on the edge of Bordeaux equipped with a carefully thought out and agreed to set of research arrangements. On the physical side, the arrangements consisted of an observation classroom in which classes would occasionally be held – often enough so that the students found them routine. The classroom was equipped with a multitude of video cameras and enough space for observers to sit unobtrusively. Other arrangements were far more complex, involving an extra teacher at each level and an agreement among the teachers, administrators and researchers setting out the responsibilities and rights of each. Nothing involving that many humans could possibly glide smoothly through the years, but the fundamental idea proved robust, and the École Michelet functioned as a rich resource for researchers for two and a half decades.

On the theoretical front, the background has roots that can be traced back through the generations, but came to the foreground in the 1960s under the title of constructivism. The title stems from the underlying tenet that knowledge is constructed in the human mind rather than absorbed by it. Applications of that tenet range from the radical constructivist belief that absolutely no information should be conveyed to students directly, to the naïve conviction that having children manipulate some physical objects that an adult can see to represent a mathematical concept will result in the children understanding the concept itself. In the interest of providing some solid research in support of the theory itself, the researchers set themselves the goal of taking some serious piece of mathematics and proving that in certain conditions the children – all the children, together – could create, understand, learn, use and love that mathematics. Accompanying that was the goal of studying the conditions themselves.

Clearly the mathematics to be used for this experiment had to be both significant and challenging. After some consideration they made a choice that will resonate with elementary teachers worldwide: fractions, or more properly, rational and decimal numbers. They had, in fact, some reservations about whether rational numbers should be taught at all, but rational numbers were firmly part of the national requirements, and likewise firmly a heavy-duty challenge for teachers and students, so they met the criteria.

The remaining element of background concerns the format for the learning adventure itself. All of the researchers were strongly committed to the Theory of Situations – in particular to the hypothesis that children could learn mathematical concepts by being put into carefully designed Situations in which they would need to construct them – but had an equally strong commitment to the principle that before people were asked to accept it they should be presented with solid research validating it. This pair of commitments helped fuel the drive to create the COREM. Once it was created, the first goal was to design research to test the theory. At the heart of that research was the curriculum that provided the adventure described in Chap. 2.

We will postpone until the next chapter a discussion of some of the mathematical choices and how they relate to the more common structure for the teaching of this topic. Our next goal is rather to set the stage for the reader to re-examine the adventure from the vantage point of the teacher. To do that, however, requires a deeper understanding of the philosophy behind the Theory of Situations and some of its practical consequences.

Public opinion in the sixties was exerting pressure for the mathematics taught in schools to resemble as much as possible, and as early as possible, the mathematics practiced and produced by mathematicians. Some even felt that from pre-school to university everything could be taught in a unique "definitive" form. However utopian the idea may appear today, at the time it didn't seem impossible to meet that challenge, or at least to study it seriously.

To do so required that the activity of mathematicians be modeled, and then that conditions be imagined that were realizable by the teacher and would lead the students to produce on their own, by a similar activity, some current mathematical knowledge. In point of fact, there is no such thing as a "mathematical activity" that does not depend on its objective, and the historical genesis of any mathematical concept is so complex and so much wrapped up in its history that it defies reproduction by any isolated modern individual. Furthermore, understanding a notion like that of rational or decimal number implies that at the end of the learning process a subject has at her disposal a collection of widely varied, logically interlinked pieces of knowledge. Thinking in terms of this end organization leads to an ordering of teaching based on logical relations, for example a locally or completely axiomatic ordering. This is the thinking that dictated the classical didactical methods.

But mathematical concepts are constructed in the course of a far richer story involving questions, problems and solutions, where a much wider collection of reasons comes into play. The researchers' idea was to realize a process of construction of rational and decimal numbers simulating that sort of genesis. That is, a process making minimal use of pieces of knowledge imported by the teacher for reasons invisible to the students. This type of project was subsequently labeled constructivist.

The initial objective of the experiment was thus an attempt to establish an "existence theorem":

- Would it be possible to produce and discuss such a process?
- Would the students – all of the students – be able to engage in it?
- Could the result of the process be, for each of the students, a state of knowledge *at least equal* to that obtained by current, standard methods?

The realization of the process made no sense unless simultaneously each lesson was conceived, studied, corrected and criticized with the most severe of theoretical, pragmatic and methodological instruments. These instruments were mostly derived from the Theory of Situations, but they were heavily modified in the course of the experiment. Another goal was thus that the instruments should progress. *The second objective was to clarify and complete the Theory of Didactical Situations.*

On the other hand, there was no question of relying on imagination and fantasy and then waiting to see if the results were satisfactory. Children are not laboratory animals. The methodological and deontological principles were very different from those in use today in that domain. In this real experiment, we set both minimal objectives in terms of success rates relative to median results at other schools, and time limits. If the method we used had not made it possible to achieve the results normally attained by classical methods in the specified amount of time, we would have had the teachers follow some alternative activities – if necessary using other methods. The comparison between two methods was thus made *on equal results on curricular objectives*, by comparing

- the time and effort required to achieve this result,
- various differences in results that were not evaluated and were often impossible to propose as objectives, of which we will speak later,
- certain qualitative differences, some of them affective: pleasure and motivation, for the students and the teacher.

The third objective was essentially to know if the use of activities similar to those of mathematicians would give the scholastic knowledge of students different qualities from that obtained by the standard teaching methods of the period.

How then were the activities set up to simulate the ways in which mathematicians generate a concept? One aspect will undoubtedly have struck the reader in the course of Chap. 2: a great deal of mathematical progress is made communally, by mathematicians bouncing ideas around, building on each others' thinking (possibly over the course of decades, but that's another story!) Correspondingly, in this curriculum a lot of class time is spent with students working together towards some mathematical goal. Given the current teaching culture, in which the expectation is that each teacher should be constantly monitoring the state of knowledge of each student and fixing up any individuals who are lagging, this can be disconcerting. A sports metaphor is perhaps the most useful tool for illuminating the situation:

How do children learn to play rugby in England (or America or Aquitaine?) Children watch the game being played and have an idea of what is going on. People run around with a funny shaped ball that if you drop it clearly doesn't do anything you want it to. After watching a bunch of kids playing for a while, a new kid asks to join in. They let him know that if he wants to be accepted he has to run in a particular direction and that he needs to throw the ball to somebody else before he gets trapped with it, that he shouldn't knock down or sock an adversary, nor cry when somebody else gets the ball. The rest of the game he learns as he plays. After a while he will even be dealing with subtleties like playing a particular position, but he doesn't need those subtleties either to enjoy the game or to make a genuine contribution to his new-found team. And if he and his team stick together for a period of time, taking on various other neighborhood teams and profiting from some low-pressure coaching, they will all learn from each other and develop both individual and team strengths. On the other hand, if someone were to break into this process and attempt at regular intervals to measure how "good" each child is at rugby, or just which skills each one has mastered, the effort would be not only futile, but damaging to the whole team's progress both individually and collectively.

In the same way, the class *does* mathematics, with everything that that necessitates and all the satisfaction it produces. Each student participates and does certain things, personally and according to her lights. What she does visibly makes a contribution to a group task, even if she doesn't fully understand every aspect of it herself. At many stages, individuals would be disconcerted and the collective rhythm would be broken if the teacher were to cut in with a form of assessment that implied that everyone ought to be able to answer some particular collection of questions. Nonetheless, as the process goes along, the whole class is developing both individual and collective understandings that lead ultimately to the knowledge in question, complex though it be.

Looking more deeply into the nature and structure of these activities requires a brief preparatory excursion into what appears at first to be a simple semantic issue, but definitely is not (it took Warfield several years to accept that it was not simple, and she is still grappling with its complexities.) In the place where the English language has a single verb: "to know" and a single noun: "knowledge", the French language has two verbs: *"savoir"* and *"connaître"* and four nouns: *"savoir"*, *"savoirs"*, *"connaissance"*, and *"connaissances"*. After numerous unsuccessful efforts to bend or dragoon the English language into conveying what we wanted it to we finally agreed that we needed simply to leave the words in French and clarify for our readers what they were saying. *Savoir,* then, deals with the kind of knowledge implied in the statement "I know that for a fact." This is not to say that each *savoir* is actually a fact. It can be a procedure, or a connection, or some other nugget of knowledge. What characterizes the knowing or the knowledge is that it is solid and certain and that it is or can be shared. In more formal terms, a *savoir* is reference knowledge. A *connaissance*, on the other hand, is more landscape than landmark. It is the feeling that the current situation is similar to a previous one whose results might be useful, or the suspicion that that tempting tactic might be a trap. It might be a little vague, or even occasionally incorrect, and furthermore it may be so unarticulated that a person is unaware of having it, but it is what gives meaning to the *savoirs*. Without a landscape, landmarks do not have much of a function.[1]

For the many occasions when this distinction is needed for understanding the issues under discussion we will use the appropriate French term. When it is not, and especially when the distinction is a distraction, we will stick with the English.

The Relationship with the Theory of Situations

With these distinctions in hand, we are equipped to take a closer look at Situations and how Guy Brousseau's theories play out in this particular curriculum. Structurally, it is easiest to think in terms of the slightly oversimplified model of a small number of general Situations in which more limited Situations are embedded (we use capital letters to distinguish these from the everyday situations that are part of the happenstance of normal life.) A general Situation would be, for instance, the exploration of commensuration that results from measuring the thickness of sheets of paper, or the exploration of the ordering of rational numbers that results from bracketing them with intervals. Such Situations are not teaching objectives, nor even problems that students must learn to recognize in order to answer them by repeating some algorithm. They are many-faceted adventures that pull together a whole conglomeration of *connaissances* that will be provoked, activated, invented, used, modified, and verified, around a project of a mathematical nature dealing with an essential mathematical notion. Within these general Situations are sequences of more limited Situations, again not focused on some specific learning objective but rather on the progression of the general adventure. Nonetheless, they are reproduced with a high enough density to be recognizable and to provoke, justify and accompany the learning, at least implicitly, of answers that suit the particular need (not necessarily immediately correct and appropriate ones.) Before long the students' answers arrive at a level of maturity such that they can be identified (recognized as stable, identical and useful), named, and sometimes made explicit by the teacher and/or the students themselves. This begins the production of *savoirs*, though at this point most are of only temporary use and value.

---

[1] This distinction is discussed further in Chap. 5.

Future learning depends strongly on the *set* of these results of an activity: the knowledge of the general Situation (the adventure) and of simpler and more identifiable Situations, *connaissances* that can be produced, improvised or manifested but only within the Situation, formulations that may be provisional and opportunistic, and of course *savoirs* that are recognized, verified, practiced, certified, detachable and exportable by analogy.

These *savoirs* constitute the only part that can be more or less formally evaluated, and as a result they tend to be regarded by some people as the only objective of teaching. Evaluation of *savoirs* alone, however, is totally inappropriate as a global instrument of evaluation and especially disastrous for making decisions about teaching (thus in particular for decisions by the teacher). To take the manifestation of these *savoirs* as the daily indicator, unique objective and unique criterion for success engages the teacher in a paradigm of extremely closed and not very productive didactical choices. Essentially it results in reproduction of the conditions of the evaluation, with a few variations and explanations to attempt to extend the useful domain of the required answer.

In reality, the teacher needs to take into account and manage the evolution of all the forms of knowledge constituting a given *connaissance*. She can only do it with powerful, attractive Situations where many different pieces of knowledge are at work at the same time, in a learning process with many repercussions, like the ones that result from the real mathematical Situations proposed here. This does not mean that learning flows "naturally" from the students' encounter with a few assignments. No Situation could possibly lead the students to the institutionalized knowledge that remains the essential, effective and contractual objective of teaching. The teacher has an on-going responsibility to keep up the level of interest of the students and the production of *connaissances* and *savoirs* of all sorts that the students themselves perceive as the results of their efforts.

What we are talking about here is a collective adventure that produces many bits of spontaneous learning that would swiftly evaporate if the process did not give the teacher and the students the possibility of unceasingly realizing the steps of a recognized didactical process. Situations do not relieve the teacher of professional responsibilities and obligations. What they provide is an opportunity for the teacher to give a meaning, a context and an objective for the knowledge the Situation gives rise to. They also allow the teacher to escape the pressures and paradoxes created by the pedagogical stance of teacher as authority and student as obedient absorber.

We have just distinguished several forms of "a" piece of knowledge. The Theory of Situations analyzes the conditions of evolution of the sets of these forms of knowledge that are at the disposition of teachers. We need to say a word about how these different sets of knowledge are determined by the position of those who are using them. The organization of *connaissances* and *savoirs* by the scientific community serves as a reference, but knowledge that the teacher wants and needs to teach is necessarily a transposed version. And cognitive psychology shows us unambiguously that student knowledge differs considerably from student to student and consequently also differs from what the teacher wants or believes himself to be teaching. Does that mean that the teacher ought to adapt himself to all those individual differences and make them the object of his work?

The videotapes of these lessons show us that the explicit object of knowledge, the one the teacher and students are working on, is the one defined by the Situation. The propositions or responses of the students are taken up only insofar as they are intelligible and useful for the advancement of the adventure. This position completely changes the relationship of the students to the knowledge that is in the course of being collectively constructed, and extends that change to all aspects of the process.

The students do not lack occasions for individually exercising their capacities. They have, in fact, more such occasions than in many traditionally taught courses. These occasions give the teacher a chance to follow the progress of the students' work without making each exercise into a blunt and decisive test calling for an immediate didactical response from the teacher. The pressure on the ones who are falling behind to catch up with the group is collective, and it is all the stronger for that.

Before we progress to the teacher's perspective on this learning adventure, let us take another, deeper look at the knowledge that the teacher is managing. In the Theory of Situations, and indeed for any thoughtful teaching, every lesson is built on various types of prior knowledge. An effective lesson modifies the knowledge, transforms it, completes it. But only a small portion of the knowledge at work in the course of a lesson attains, by the end of that particular lesson, a state that permits the students to formulate it and fully understand it (and thus to be able to write it down as a response to a standardized (decontextualized) question).

In general, before it can emerge as a *savoir* and be exported out of the situations in which it has made its original appearance, knowledge must progress as a *connaissance* in hidden forms through different lessons, often numerous and widely dispersed. A *connaissance* is initially tightly attached to specific situations and limited by the role it plays in those situations. To be detached from them and take its place as a *savoir* it must be recognized, formulated and analyzed. That can be a long process, one that constitutes a genesis of that *savoir*. In every lesson several notions are under construction, often in different stages. Thus the teacher manages (teaches, provokes, sustains, rectifies, etc.) a whole bundle of different *connaissances* and *savoirs* in varying stages of development. The means of managing each one is a Situation – or rather the role that a Situation makes that knowledge play by provoking or justifying its use, its transformation or its replacement. The teacher must thus add or deepen Situations and the means of resolving them and also find within them the questions that keep the process unfolding.

On the other hand, at a given moment, even if the Situation being worked on as well as the knowledge needed to resolve it are common to all the students, the relationships that individual students have with the Situation and the knowledge are all different. The maturing of a piece of knowledge is frequently spread over several lessons. The behaviors of the other students form part of the didactical Situation, and consequently it is not possible to synchronize all of the didactical events among all of the students. At any given moment the teacher must be able to deal with leftover, undigested bits and forms of knowledge as well as newly arising ones. That does not mean that she needs to prolong the process in order to keep addressing the old forms, but that she must not make it impossible to progress if the knowledge is still a bit imperfect. In order to do that, she must constantly assess both the state of knowledge of the class as a whole and that of each individual student. This provides a totally different kind of information from that provided by an examination using a pre-set collection of questions. With this information she is able to make and carry out a continuous sequence of didactical decisions.

The teacher is dealing at the same time, without confusing them, with class knowledge and each student's knowledge. These are different forms of knowledge, and are differently manifested. Hence the knowledge that reaches maturity in the course of a given lesson does not show up in the same way for all of the students. The process must make it possible for the knowledge that is indispensable for community use to be shared as swiftly as possible by the whole class, while leaving some leeway for less immediately crucial knowledge to be developed at different paces by different students.

To consider the objectives and results of a lesson exclusively from the point of view of certain *savoirs*, focusing especially on which ones have not been acquired (which in effect is the normal tactic) is insufficient for managing and conducting a learning process and in the long run dangerous. The minimal objective of a lesson should be to make it possible to approach the next lesson in good condition. The results of a lesson are represented by the number of lessons that can be taken up after doing it that couldn't have been taken up if it had not been done.

A particularly clear illustration of a Situation where class knowledge and individual knowledge tend to diverge and require a lot of managing is the sequence in which the decimal numbers are motivated and introduced by using intervals to bracket a fraction [Modules 4 and 5], about which there will be further discussion later in the chapter. These lessons make unusually heavy use of class knowledge as distinct from individual knowledge. Certainly by the end of the sequence, the individual knowledge of all (or essentially all) of the class includes the forms and uses and management of decimal numbers, and furthermore a well internalized notion that they resolve some messy problems with rational numbers. On the other hand, at many of the intermediate stages the process depends almost exclusively on a more general form of shared knowledge, where everyone is engaged, and everyone has enough partial knowledge to play a genuine part in developing the Situation, but very few if any have the whole picture in their heads. The results in terms of depth of conceptual understanding are well worth the effort, but there is no denying that the process is extremely challenging for the teacher!

The Perspective of the Teacher

Let us move on, then, to the perspective of the teacher. The adventure of these students was also – and above all – that of the teacher. What decisions did he need to make, based on what indications? Our look at the adventure from the student perspective does not tell in what ways the teacher was free to adapt his lessons to the results of the students. There seems to be a great discrepancy between the complexity of the lessons and knowledge that the teacher was responsible for and the apparent simplicity of the knowledge – that of an ordinary class – ultimately provided and formally verified. What is the meaning of phrases like "All the students took part in the activity and finished it", "The students understood that …" or, "After that the students knew that …"? What was the final result? Why do there seem to be so few formal "learning exercises"? It is the adventure of the teacher that we will try to describe here to respond to these questions.

The teachers who had to manage this curriculum had a high density of aid from a team of researchers and advisors who explained the design, tried to understand the difficulties encountered, and attempted to respond to them. The teachers took part in figuring out the concepts their advisors were using and understood them very well. We will be speaking of the teachers here in their role as instruments of the work. But these teachers were solely responsible for what the students did. They had not only the right but the duty to refuse any suggestion that seemed to them not to be good for the students, and to put an immediate stop to any activity that got out of their control. Very swiftly, by reproducing the same curriculum each year, they familiarized themselves with the profound modifications required in the ways of managing class, and adapted themselves marvelously to it. This is why, in this chapter where we want to look at the adventure from the spontaneous point of view of the teachers, we must anticipate the following chapter and mention some theoretical concepts.

In circumstances where testing plays a heavy role in the evaluation of teachers, schools and even the whole system, teachers are under pressure to focus on results that can be observed by means of standardized tests. Most of their decisions then depend on this ultimate step of the teaching process, and most of the techniques that are considered acceptable are based on the corresponding type of formalized reference knowledge, or *savoir*. The present curriculum offers an alternative by working with all of the *connaissances* – general knowledge in all its forms and stages of development – that preceed and accompany *savoirs* without themselves being either *savoirs* or scholastic objectives. These *connaissances* are picked up in encounters and dealings with appropriate situations. They play the same role that the family environment plays in the learning of native language.

In the course of the process of teaching that we are presenting, a *connaissance* evolves and changes form, use and meaning. In this way it becomes more precise and complete and ends up being known in the canonical form that the culture assigns it, as a *savoir*. This *savoir* results primarily from living with these *connaissances* in many forms. All of them contribute more or less to the moment when it is suddenly obvious that "Everybody knows that…". Knowing how to recite the rules of the road requires much more effort and is less effective than knowing them because one has practiced them assiduously and knows the reasons for them.

The success of each step depends on the previous ones and more or less conditions the possibilities for the ones after it. The collection of these steps constitutes the process of teaching and learning of a *savoir*. In the course of each step, a number of *connaissances* are engaged, each at a different stage of development and evolving towards a different *savoir*. The same *connaissance* presents itself in diverse forms: decision, formulation, explanation, which appear and evolve in appropriate situations.

The teacher does not evaluate *connaissances* like *savoirs*: it is how the activity itself works out that indicates how the project is advancing. The importance of having each student participate and do what she has to do is not the result of an abstract educational intention, it is a necessity – like talking, sharing a culture or a project – a piece of evidence for the students and the teacher. That way the students participate in the development of the curriculum. That is what they call "doing mathematics" as opposed to "learning mathematics" (which they also need to do at times).

Carrying out such a process is at once more complex, more demanding of the teacher in terms of engagement, and less alarming for all parties.

A metaphor might help: the teacher braids a rope whose strands are evolving *connaissances*. A particular *connaissance* may appear and develop and wind itself in with some other *connaissances,* then disappear from sight, only to reappear further along the rope as a new strand that develops in perhaps a different direction and winds itself in with yet another set of *connaissances.* The thicker the resulting rope, the stronger the knowledge that it represents.

The art of the teacher resides in the possibility of observing each stage of the progression of the curriculum and associating with it the decisions most favorable to the stages that follow. Sensitive observations and reliable models for decisions are essential conditions for obtaining chains of decisions – though not the only conditions.

We will first turn our attention to the basic question:

How does the teacher manage the progression of the Situations and the learning of the whole class? How does she continually assess each student's behavior along the way towards appropriating some mathematical concept, and how does she deal with possible divergences from the intentions of the curriculum?

The accuracy of the curriculum and the intimate knowledge of it that the teacher acquires in successive reproductions of it are helpful and reassuring. But a closer look reveals a wide array of possible accidents, detours and divergences. The success of the curriculum and of the students is a result of constant vigilance over certain variables, of constant exercise of subtle choices of judicious decisions, and of clever corrections to prevent the students from losing interest, scattering and giving up.

For simplification, the teacher distinguishes four major types of lessons:

1. Lessons introducing a concept

   These are lessons that introduce the students to an important new mathematical notion: the Thickness of a Sheet of Paper (Module 1, Lesson 1), the Puzzle (Module 8, Lesson 1), the Enlargement of the Optimist (Module 9, Lesson 1), the Pantograph (Module 14, Lesson 1). These lessons are fundamental ones, which we were able to conceive in such a way that they almost invariably produce the desired behavior from the students. The role of the teacher is far from negligible, but it consists entirely of predicting and preventing any accident from messing up or slowing down the dynamic of the game, of directing the didactical phases with spirit and conviction, of discretely encouraging perseverance on the part of some whose energy is flagging, of welcoming student involvement with interest even when it is slightly off track and leaving the Situation to make any necessary corrections to these indispensable contributions.

2. Intermediate Lessons

The students invest their fresh, new knowledge in intermediate Situations to solve known problems in slightly unfamiliar conditions. The teachers found these lessons to be the easiest to manage. It is always a matter of resolving a mathematical question. The students discover new knowledge in problems that make them use and use again the reasoning and calculations that are becoming familiar, but are not yet frozen into scholarly conventions.

For example, Module 8, Lesson 4, the Image of a Decimal, is a typical transition lesson. The introduction of the Situation appears to be the same as that of the Puzzle which immediately precedes it, but the measurements are in decimal numbers, not integers. The children have just finished constructing decimal numbers as a means of comparing and ordering fractions, but these decimal numbers are not yet objects of *savoir*, directly usable in a canonical manner. Sometimes they function almost like whole numbers (for ordering and operations), sometimes the students have to go back to their fractional form to figure out their still somewhat astonishing behavior. The teacher has not yet established one of the different modes of calculating fractions as a canonical method, which would have transformed the whole Situation into an exercise.

The proposed Situation, like that of the Puzzle, has the capacity to reject a fair number of the incorrect answers without the teacher having to intervene. On the other hand, its mathematical objective is considerably more modest than that of the Puzzle, which is designed to produce the discovery of a whole new property.

The students work in groups of three, but each student has the responsibility of producing a piece identical to that of his neighbors, which must fit with theirs to produce a tessellation. This task gives rise to observations that are not an objective of the sequence, but do serve to maintain the interest of the students. For example, some of the groups set about to calculate the eight segments of the perimeter independently, but observation of symmetries enables others to see that they can get by with just three calculations. They point it out to their teammates, which brings out some questions and explanations.

The teacher circulates among the desks and observes the progress of the operations. She might intervene if something of no specific importance interferes with the work of the students, but not to suggest or correct the reasoning or realizations of the students. Only if an error is manifestly sterile, blocking, and incapable of fulfilling its role of pointing students in the right direction does she step in.

Decimal measurement to the nearest millimeter is one of the results of a preceding phase that is built on 3 years of familiarity with the ruler. But it is not an objective of this lesson, especially since an error in the measurement of the model would only surface very late in the process. The teacher has two students measure the sides and write their measurements on the board for everyone to use. The work of the students deals with the method of calculation and the calculation itself. The teacher insists on having students carry out the calculations individually before comparing the results with others in the group. But it is acceptable to help a comrade with one or two of the calculations, and to discuss which method to choose. There is absolutely no obligation for all the members of the group to use the same method. However, a member of a group is allowed to insist on understanding what another member did.

Students are thus motivated to use methods that others can understand or at least reproduce so as to make their own calculations. But they are also motivated to have several methods available if possible, so as to present them for their classmates and the teacher to admire in the presentation phase of the lesson. The teacher permits this goal but doesn't encourage it much, in order to avoid the proliferation of equivocal propositions that could soak up a lot of everybody's time and energy.

No explicit reference is made to the procedures used in the previous lesson. The students do not have to reproduce what they did the day before, just to use it for inspiration without losing sight of what they are now doing. That makes an implicit rule for the teacher, who must avoid saying, for instance, "Just do what we did yesterday!" That would be a purely didactical argument. This Situation is different and should officially be examined independently. The similarities are the student's responsibility. To be sure, the expectation is that the student will use or try to use what he did the day before, but of his own volition.

Since the numbers are the same for everybody, it is hard to maintain the uncertainty. It is absolutely necessary that the individual part be respected and be required for the making of the pieces. The teacher needs to verify that each student has had to carry out by herself some calculations similar to the ones from the lesson before. If it is needed, the teacher gives different groups projects with different dimensions.

This lesson is close to being a classical exercise. The children do carry out similar calculations over and over, but here it is in a completely different spirit. These calculations are justified by a collective task, not by a personal project of perfectionism required by a monitor. Knowledge is made evident by its use in a new "adventure"; it is going to become familiar, with or without the aid of formal description, which will not turn up until it is needed for the development of further knowledge. In this process, the pressure to turn the scholarly activity into an individual formal learning project is minimal. The engine is the participation in the construction of a collective and individual culture.

3. Terminal lessons

The following lesson (Module 8, Lesson 5) proceeds just the same way, but it is a different type: it provides a conclusion and an institutionalization. It looks like a continuation of the preceding one – it takes up the same *miliou* and it is still about a fixed enlargement: $1 \rightarrow 3.5$. But the questions are very different and not "motivated". The teacher asks for the images of a bunch of numbers that are clearly of a particular kind: 1/10, 1/100, 1/1,000, etc. In the process of carrying out the calculations, the students come to the realization that they can now deduce from what they already know a new (for them) rule for division of a decimal number by 10, 100, 1,000, and that they can say it, prove it, practice it on demand, and require other people to understand them without having to repeat their demonstrations It is just a question of recognizing what they already know how to do and nailing it down with rules and words that express what they already think and know. The numerous calculations that they have to make are justified by a goal held by the community of students, but it is a knowledge goal that they can see approaching and that they achieve.

This looks exactly like a classical lesson, except that it is the students who are supposed to guess and establish the *savoir* that is to be learned in order to resolve the situation proposed. It would be only an exercise if the method had been laid out in advance. It is completely simple to solve, and the students work individually. The question is different and arbitrary, but the answer is known (not for sure, but it can be guessed.)

The numerous individual attempts are not repeated exercises. They are attempts, more or less successful and more or less appreciated by the others. The goal is to be able to continue taking part in the common adventure with the other students, to be able to present one's ideas and bring in one's work. It is not the pursuit of a personal egocentric goal supported only by the undependable satisfaction of adults.

The formulation of the rule for dividing decimal numbers by 10, 100 and 1,000 is stated by the students at the request of the teacher, accepted (i.e., institutionalized) as a *savoir* and immediately applied in exercises that are promptly corrected. This is the normal method, and it has the usual results. Many of the students understand, all of them make some correct calculations and many make mistakes. The teacher is not expected to hold out for an immediate, definitive, and general success on this important question. Because it is used frequently, they will be reminded of it often and the teacher can follow the individual progress of the students until they get it. The goal of Situations of institutionalization is for the students to know that they have a common repertoire of objects, terms and *savoirs*, which can be best understood in exchanges with others if they use the conventional solutions, terms and explanations.

4. The process of generating a concept

The most complex lessons for the teacher are those where for an extended period she must manage provisional, uncertain knowledge in order to bring out different aspects of a concept. Ambiguities are only gradually resolved, nothing is formalized but nothing should be forgotten.

The best example of this type of process is the sequence of Situations leading up to the construction of the decimal numbers (Module 4, from Lesson 1 to Lesson 4). In this type of sequence the teacher and the children use and evolve *connaissances* that cannot be set up as *savoirs*. Every Situation prepares for the one that follows as much by the questions it raises as by the answers it provides. The most important thing for the children is remembering not the specific outcomes of the adventure but the things they have encountered along the way – intervals, end-points, interval lengths, the search for a strategy for reducing the interval of uncertainty, etc. Nothing is to be learned in final form, but all the calculations they have made contribute to an incomparable familiarity with the rational and decimal number line, and with the calculation and location of those numbers.

These lessons have to do with the order and topology of the rational and decimal numbers. They come close to reproducing an almost historic and scientific development, but the objectives and real significance of the sequence remain obscure to the students until quite late. It's a matter of comparing the size of fractions, finding an interval around them, estimating them, ordering them, improving on the intervals, and the like. At the end of the route, after Module 8, Lesson 5, the students invent a method that could be called a division, but that for them is just the method of finding a decimal expression for a fraction by locating it in successively smaller decimal number intervals.

In the opinion of the teachers, the first lessons of this sequence were the most difficult ones to manage in the whole curriculum. Nonetheless, they were successfully reproduced every year for 25 years with the same results.

Within the lessons, whatever the type, the teacher must make choices based on the state of knowledge of the students, which brings us to our next question:

What are the manifestations of student mathematical activity with respect to a connaissance?

In the course of carrying out a Situation, the teacher must keep track of the functioning and evolution of many forms of *connaissances* related to the *savoirs* that she wants her students to acquire.

A major mathematical *connaissance* makes its appearance in the curriculum as an initiative of the student in different roles and conditions, and generally roughly in the following forms and order:

Observable Aspects of Connaissances

- Student decisions. For these the *connaissances* need only be adequate for decision making, whatever the form in which they are conceived (Action Situations)
- Formulations that may be improvised but must be intelligible (Formulation or Communication Situations)
- Proofs that it are valid, and consistent with what is already known. The proofs must be recognized as valid by peers (Validation Situations)
- *Savoirs* extracted from their context and offered in a situation where there may be doubts about their pertinence or utility, but not about their validity.

Savoirs follow a different route, since their status as reference knowledge needs direct action from the teacher. They nonetheless need to be kept track of

Manifestation of Savoirs

- As a reference: its definition or certain of its properties, expressed in a canonical fashion, are declared or recognized by the teacher as personal, interpersonal or cultural references (Institutionalization Situations)
- Explicit investment of these references by the students in problems or exercises and in proofs.
- Casual use as references or implicit knowledge in new uncertain situations (Action Situations)

Although the succession is not arbitrary, it is also not a formal necessity. It can be adapted and be responsive to the possibilities and necessities of the curriculum, which itself is subject to many other constraints. In this curriculum, all unnecessary steps and digressions have been eliminated. Some Situations produce a rapid evolution while in other cases several Situations are necessary to achieve a single step. Different *connaissances* are involved in the same Situation, in different forms and roles. They may advance all together or separate themselves in conjectures or reasoning. This process simulates as authentically as possible a genuine mathematical activity.

We will not detail here the tangle of *connaissances* and *savoirs* that turn up together in the course of each lesson, each evolving in a specific way under the influence of successive Situations in the course of the curriculum. The teacher must stay conscious of the dependences that come into play among these *connaissances* in the course of the different steps. The reader can follow the twists and turns of the adventure in Chap. 2. Here the issue is to understand the action of the teacher while the adventure is in progress.

Teaching a mathematical subject presents a teacher with two essential and distinct types of difficulties: those connected to carrying out each episode (a whole lesson or a particular phase: assignment, exercise, correction, assessment, etc.) and those connected to the total trajectory: choice of successive episodes and the passage from one episode to another (or from one lesson to another.) The former have to do with the actual activities of the students moment by moment, and the latter with the possibility that these activities can succeed in producing a coherent culture, and a capacity among the students for undertaking new activities. Concretely, in the second case, for the teacher it is a matter of evaluating the possibility of undertaking the next phases of the curriculum based on the earlier ones.

In terms of Situations, the result of a particular episode consists of all the Situations that can be taken up thereafter with a good chance of success but could not have been before it, and of all the ones that will not have to be revisited at the end of the teaching sequence thanks to having done it.

In the curriculum that we are presenting, the principal instrument of regulation at the disposition of the teacher is the choice of the moment of institutionalization. In supporting autonomous activities of the students, the Situations bring out questions, convictions, declarations, arguments, *connaissances* that are justified only by their temporary use in the students' thought process and in this particular Situation. The cultural value of these *connaissances* – their actual validity, their canonical formulation, their place among *savoirs* – is not something the students can deduce from their role in the Situation. Furthermore in the course of the Situation events turn up that are known to only one student or group of students or even to the whole class, and the students don't know their value and may suspect that they are temporary, since the Situation itself may modify them.

Institutionalization is an act or process that causes a fact or *connaissance* to pass from one sphere to a larger one. For example, the teacher tells the whole class about something done by a student or group of students, or summarizes the session from the day before and the state of the question being studied, or describes a result that everyone can now count on, or confirms that a conclusion conforms to the truth and is recognized by science or society, indicates that a result was the objective of the lesson, etc.

Institutionalization has a slightly ambiguous status. The *connaissance* or convention in question is certainly precise and well determined, but the affirmation that everyone will henceforth know it, practice it without opposition, and use it as means and reference is a convention and above all a gamble. The fact that no one is supposed to be ignorant of a law does not turn that law into a sure practice. The fact that not everyone respects or is able to respect the law is not a reason to give it up.

Institutionalization of *connaissances* is a Situation in the course of which the teacher recognizes as valid and accepted by society the *connaissances* that the students have come up with in the course or conclusion of a Situation or series of Situations and that they propose as a reference. This event concludes the phase of quest for *connaissances* on the part of the students and determines the *savoirs* that they can take as certain.

Normally institutionalization signifies that thereafter each student will be authorized to refer to this *savoir* to support an opinion, and is assumed to be capable of producing it with precision and confidence and using it correctly. The teacher guarantees that this *savoir* is exportable and recognized outside of the classroom by society as a whole. Clearly nobody in the class but the teacher can give this guarantee.

So the question is how to determine the moment at which institutionalization can be made to have the best chance of succeeding with all of the students. Done prematurely and suddenly it would isolate the few students who were the first to be able to understand it and submit to it and would tend to make the rest appear to be rebelling against a communal law. Not only that, it would make the latter submit to a servile relationship with *savoir*, to learn and apply a rule that they do not understand and that they can only acquire by procedures foreign to their understanding. At the other end of the scale, an excessively scrupulous institutionalization would wait until each and every student understood and could put the rule into practice. Waiting that long would cause an excessive delay in pursuit of acculturation to other *savoirs*.

Institutionalization can apply to *connaissances*, but also to Situations. When the development of the Situation becomes confused, the reactions and the various more or less true or false "*connaissances*" diverge. Nothing more can be understood the same way by the whole class in the natural course of the actual Situation. These differences make the pursuit of the proposed communal activity impossible. The teacher must then pull everyone together with "What has happened so far? What was the Situation we started with? What did some of you do? What did others of you do? Where are we now with the problem?" This re-framing of the Situation informs all of the students what is in question, what deserves to be noticed and what remains the object of the action, which can then resume its course (unless the essential part of what was of interest in the Situation has had to be revealed).

This approach to institutionalization contrasts sharply with the curricula (such as the daisy-chain programs) that are reduced to a sequence of institutionalizations. Each lesson, each exercise and each *savoir* presented is considered to be both necessary and sufficient for proceeding to the following step. Every question is considered to be equally key and definitive and the only *connaissances* considered are *savoirs* and errors. At every step the student is supposed to make an effort sufficient to succeed in completely acquiring a given, indispensable *savoir*.

Institutionalization marks the separation between things that are of the order of *connaissances* – temporary, personal, in question – and things that are accepted as definitive, agreed to, common and sure.

In making decisions, teachers must be conscious of the whole structure of the curriculum, which brings up the question: What are the dependencies between lessons and between things learned?

How does the progression of one lesson depend on that of the preceding lessons? What are the indicators of good progress in the process? What are the possibilities for intervention by the teacher if something goes off track or fails? What constitutes a failure and what is just an episode? All these questions are tightly linked.

How a lesson develops can depend on how the previous one developed. The second one can depend on the *savoirs* learned in the course of the first. Sometimes the students cannot do, say, understand or learn what they are supposed to because they cannot use the necessary *savoir* because they did not learn it beforehand.

The precaution of never using a single word or property that has not previously been defined or demonstrated is the basis of the general, deductive organization of mathematics. This organization is often used as a model for the teaching of science and even for the acquisition of all scholarly knowledge. The teacher wants to be able to report that he has made available to the students all the necessary elements and the only possible cause for failure of his lesson would be failure of previous teaching or inability of the students to understand the construction under way. But before *connaissances* can be definitively cast in the bronze of organized *savoir* they must be established by complex processes very different from this fiction. In this curriculum we try to have the students reproduce or simulate such a process.

To this end, we have installed at the heart of the lessons Situations that are steps in an adventure. A Situation exists independently of the actions and modifications of the protagonists.

Two successive Situations are linked if what is produced in the first conditions what can be produced in the second. We distinguish at least two types of dependence:

1. Two lessons may be linked because the second (in time) uses or resumes use of *connaissances* that have been *established* in the first. They are connected by a structural relationship of *savoir*: for example the second lesson studies the corollaries of a statement established in the first.

   The reality of the learning sequence does not necessarily follow the order of an exposition of *savoirs*. It is not indispensible for the students to have fully understood and learned everything that has been defined or demonstrated for them. The study of the consequences, extensions and "uses" of taught *savoir* is indispensible to explore, know and understand a definition (which furthermore is often the result of a concentrated result of a complex process.) Otherwise stated, the appropriation of a *connaissance*, even presented in a strict axiomatic order, depends as much on the lessons that follow it as those that preceded it. The process must be considered as a whole.

2. Two lessons can also be linked not by the relationship of the *savoirs* that they offer for learning, but by the questions that these *savoirs* are supposed to answer. The result of an unresolved Situation can be a new question that gives rise to a new Situation which itself may make it possible to resolve the initial problem.[2]
3. A lesson can obviously be linked to a previous one both by questions arising from the previous one and by the consequences of the *connaissances* established in it.

For example, when the teacher asks, "Are the measurements of thickness numbers?" he introduces a natural sequence of lessons (i.e., one that the students could practically run themselves) sparked by the questions "What do we do with the natural numbers that we may or may not be able to do with the measurements of thickness?" And certainly also new *connaissances* are established using the preceding ones.

What Then Are the Causes of Learning and the Reasons for Knowing?

Having an individual reproduce the same task is the antique means of having him learn it and execute it more easily in all circumstances. The learning can be observed through the progress of the student in the perfection of execution (reliability, speed, precision). The link between the successive steps is essentially the state of the student. One cannot pass from one task to a more complex one unless the student has satisfied certain required conditions. If there is a connection between the things learned, a progression in the complexity of the tasks, only the teacher responds to it. Thus it is the state of the student that is the link between two steps of the learning process. The student reproduces calculations in order to know how to do them. And if learning makes no progress he has only himself to blame, his characteristics, qualities or faults. The teacher and society reinforce the blame and question the properties and virtues of the individuals who are recipients of the keys to perfect learning.

The learning process with which we were experimenting here is completely different. This formal (and universal) learning process has only a marginal place in it. The repetitions of "exercises" are not motivated by a direct desire to enrich oneself by knowing how to do them. They are steps in the realization of a task that has its own significance and interest and that is a goal shared with others. The Situations proposed are not solely destined to be the causes for learning for individuals, they are first of all destined to determine the reason for some *savoir* to exist, the role that it plays in people's relationship with each other and the world, and the role that humans play in society thanks to that *savoir*.

---

[2] An experiment that we carried out demonstrated how a sequence of Situations each issuing from the previous ones by questions produced by the students was able to generate the discovery of limits of frequencies and of measures of events without the teacher's ever proposing a Situation beyond the initial one or supplying information or a personal solution – and without the notion of chance ever being mentioned! (Brousseau, Brousseau, & Warfield, 2002)

There are large moral, cultural and epistemological differences between reproducing a calculation in order to advance a common task and reproducing a calculation simply to know how to do it oneself.

Finally, the conduct of the lessons and of the sequences they formed depended heavily on the observation by the teachers of a certain collection of indicators, on the verification that a certain number of appropriate and expected corrections were resulting from the combined impact of the Situations and the teacher's interventions in these Situations.

How Does the Teacher Use Assessment of and Within the Curriculum?

The teacher assesses the Situation under way, the *savoirs* in action and the students. These assessments are subordinate to the possibilities for action that are available to the teacher and depend on the assessment.

The purpose of assessing the Situation is to determine moment by moment whether it is best to let the Situation proceed or whether it is time to intervene and either to redirect its course or to interrupt it. For example: Is some additional commentary on the assignment needed so that all of the students have some project for action (whatever it may be) that will let them get into the problem? On the other hand, at what point would supplementary information make the necessary efforts useless?

The decision depends on the expected profit from the amount of additional time accorded to the intervention. It is difficult to describe in a few lines all of the factors that need taking into account: the fatigue or loss of interest of the students, the amount of useful information that can be harvested (not just plain success.) Sometimes teachers content themselves with one correct proposition (the success of one student or group). Sometimes it is important that each participant obtain a proposition to present to the other students.

This assessment applies simultaneously to reality – to facts – and to their meaning, that is, the possibilities for interpreting them offered by the Situation. Sometimes making each and every student experience all of the difficulties and their solutions is completely superfluous. Students may be able to benefit by proxy from the experience of others. Sometime simulations are sufficient, while other events need to be really experienced. Deciding to organize a Situation in such a way as to produce the discovery of the properties of a mathematical notion is a non-trivial decision. It takes what might be a considerable amount of time for the sake of what might be a trivial significance. Any time that that is possible the Situation needs to be reduced. Often a short definition followed by an illustration of examples and counterexamples is the best solution if that definition is useful in the project in progress. Often a simulation can be worth more than an actual heavy realization.

In general, Situations define and at the same time more or less dissemble certain *connaissances* that the student is supposed to make use of to accomplish a proposed goal. Certain Situations have the objective of determining whether the student has available, directly, the *connaissance* for the solution. By definition they do not offer the students who do not already have this *connaissance* the possibility of answering. They are assessments for the student, but they are thus not in principle didactical. They answer a question, but say nothing about others that might connect to it.

Others, on the other hand, have the (didactical) property of inducing the production by the student of a *connaissance* that he did not previously have available (in the form of a *savoir*), but that he can conceive (guess, construct, comprehend, etc.), formulate, prove valid and finally "learn" at his own pace.

When a teacher must intervene in order to promote the evolution of a didactical Situation, one of the principal difficulties consists of monitoring the informative value of his interventions. In an effort to stimulate or re-launch the activity of the students he might bring in information that reduces the Situation to the obvious, or on the other hand he might complicate the work of the students by throwing in superfluous intentions or requirements.

Didactique is, for the teacher, the art of showing and hiding his intentions in such a way as to permit the student to discover as a personal response to objective conditions the thing that the teacher wants to teach but cannot reveal without depriving the student of the possibility of doing it himself.

Making *connaissances* contribute to the learning of *savoirs* so as to approximate the real cultural, social and psychological functioning of mathematical thought presents some very real risks: first, the risk of wasting time and energy, next the risk of accidently producing the learning of *connaissances* that are false, or badly established, or badly formulated, inappropriate or culturally unacceptable.

It is very important to know how to interrupt a Situation that is becoming ambiguous, or that doesn't guarantee that the emerging *connaissances* will have a reasonably strong and simple impact. One must not hesitate then simply to state the canonical solution being sought. This danger is eliminated in the curricula that only consider established *savoirs*, as visible objectives and/or as means.

The Assessment of Students and Groups of Students

The goal of assessing students is to predict whether they are going to be able, together or individually, to take on the rest of the curriculum. This is of interest exclusively in the case where it is possible to choose and manage the curriculum on the basis of the results of the assessment.

Naturally the progression of the Situations permits the teacher to adapt a Situation to the possibilities and varied talents of the students. This continuous adaptation is easier than the choice of appropriate exercises and problems. But once a Situation has come and gone, once an adventure has been lived, for better or for worse it can't begin again. Moreover, the construction of *connaissances* and their meanings is common to all, and there is no royal road.

It is at the moment of institutionalization that it seems one must discriminate between those who understand and those who have not gained from the curriculum the resources needed for the proper acquisition of *savoirs*. Then all that is left is classical resources, explanation and repetition. These possibilities must not be neglected. Institutionalization does not put an end to the process of learning. Really useful *connaissances* should be revisited often enough to permit a party of students to rejoin the troop. What a Situation has made the students live, what has been perceived, communicated and explained is not a required object of *savoir*. People from the same society live and communicate with highly varied repertoires.

The Types of Situations That Appear in the Lessons

We made a distinction above among types of lessons, distinguishing them by their function in different stages of the learning process. Another perspective on the teacher's role comes from a dichotomy that is deeply rooted in the Theory of Situations: *didactical Situations* and *a-didactical Situations*.

- In *didactical Situations*, the teacher maintains direct responsibility for all stages of the lesson. She tells the students her intentions, what they will have to do, and what the results should be. She intervenes freely to keep the class traveling on the desired route. In our curriculum the reader can spot these completely classical phases. They were carried out in the classical manner.
- In *a-didactical Situations* it is the students who have the initiative and the responsibility for what comes of the Situation. The teacher thus delegates part of the care for justifying, channeling and correcting the students' decisions to a *milieu* (a problem statement, a physical set-up, a game, an experiment).

The former tend to produce the learning of reference knowledge, either permanent (*savoirs*) or temporary (assignments, rules, etc.). The latter tend to bring into play *connaissances* corresponding to the *savoirs* being taught. *Connaissances* manifest themselves in responses (actions, choices, expressions, trying things out) in circumstances where they seem necessary and adequate.

In didactical Situations, the students' *connaissances* do not develop and are only manifested in the course of applications, and thus after the presentation and acquisition of the necessary *savoirs*. The teacher demonstrates that the expected answer has been given in the preliminaries, or convinces the student that it is his responsibility to deduce it from what he has been given. But in fact *connaissances* can appear before the student has the corresponding *savoir* available in appropriate circumstances. Thus it is possible not just for *connaissances* to follow from the acquisition of *savoirs* but for them to precede and justify that acquisition.

These *connaissances* correspond to a *savoir*, but they may well differ from it (for example they may be true or false, or consist of beliefs, or be questions.) They may also differ from student to student, because they are often individual. They are similar in the sense that they tend to be opportune and adequate in the same circumstances. They are fleeting and cannot be directly evaluated by classical tests. But they are the only means by which students can participate in the adventure of their own learning.

Classical curricula also combine didactical and a-didactical phases, but in this curriculum their roles and relationships and the proportion of time allocated to one or the other are profoundly modified. Our aim was that the *connaissances* be also the means by which the students participated in the epistemological adventure of *savoir* and of their own personal *savoir*. The curriculum presented in Chap. 2 develops *connaissances* that precede, accompany and follow *savoirs*, as happens in the natural exercise of mathematics.

But this ambition complicates the work of the teachers a lot. What is that work, then?

The Types of Didactical Situation and How They Are Conducted

Many didactical Situations were classical and were (and are) in use in all schools. But the reader may also note some didactical Situations of a new type:

Situations of Institutionalization

These were discussed above, but we will expand slightly on them[3]: The teacher directs a session that consists of observing that almost all the students understand *this* and know how to do *that*. He has the students put this *savoir* in order by presenting it himself. He makes the definitions, algorithms and theorems precise and declares that henceforth he is counting on the few students who are still hesitant to look into these questions in order to be able to continue to work with the others. These are the didactical Situations in which the students learn that certain of the *connaissances* that they formed in the course of preceding a-didactical Situations can be organized, formulated and thus proved. They learn that henceforth they need to know them for communication and for reference. These are lessons of institutionalization. They take on a particular importance because of the importance given to the a-didactical Situations for developing *connaissances* before putting them in definitive form.

These lessons are delicate. Only the teacher can judge the best moment to activate this phase of learning. If it is done too late the children will have developed and become entrenched in ill-conceived ways of doing things, inappropriate ways of saying things and fallacious reasons for knowing things – *connaissances* ensconced as *savoirs*, but badly built and difficult to abandon. If it is done too early the *connaissances* will not be sufficiently familiar to support a precise and solid formalization. A large majority of the students must be able to make the change without effort in order for the

---

[3] See also Chap. 5.

challenge to be met by those who need to make use of the following encounters with this *savoir* to finish learning it and need to do some exercises to make it familiar.

Institutionalization in this case deals with *connaissances* and produces *savoirs* that are durable and if possible definitive. It can also deal with provisional conditions such as the rules of a game. It is often easier to learn the rules of a game while playing it than it is to learn them in advance. The operation is only interesting if the rules are simple, if the game is reasonably easy to repeat, if the student can notice for herself the causes of any difficulties and correct them and if the *connaissances* thus produced are both interesting and useful for the learning being aimed at. In this case, the rules are part of the solution *savoir*.

Situations of Devolution of an A-didactical Situation

Students are only willing to enter into an a-didactical Situation in the hope of finding pleasure and profit. They must have the hope that they will be able to find on their own the essential parts of the solution, and that the search itself will be exciting and intriguing, that it will be reproducible (though an occasional serendipitous victory produces a kind of satisfaction and should be accepted.) Otherwise stated, the "games" chosen must present specific real qualities and notably feed-back that permits the student to check the value of her actions and understand the reasons for it.

This does involve a didactical Situation because the teacher must teach the rules, but his role consists principally of indicating to the student that he has no obligation to tell her what he wants to teach her. The teacher must let the student know that he ardently hopes she will play, but he cannot force her to do so, and that he hopes not so much that she will win as that she will understand and learn something that will enable her to win.

Conducting such a lesson is a difficult art. The teacher must show a great interest in the game itself and give encouragement to all the players, but he must respond indistinguishably to successes and to errors or stupidities, and initially treat discoveries as difficulties. It is the students who must judge what is good to know, true and useful. The teacher must be able to encourage the students and help the weak or the suspicious – but not too much.

These Situations are not made for judging the students but for developing and judging *connaissances*. For that the teacher must supervise numerous parameters: what it costs the students to participate, the speed of their progress, how ideas spread through the class. He must calm fears and also excessive enthusiasm. If the Situation is not well calibrated he will have to make concessions, but he will have to hide them as much as he can.

The a-didactical Situations that permit this devolution cannot be improvisations.

Situations of Evaluation

In the traditional system what is evaluated is essentially the students, indirectly the knowledge acquired, and secondarily the teachers. But this type of summative evaluation gives only partial information, insufficient for making decisions in the course of learning. This information is subject to superficial interpretations produced by reductionist pedagogical ideologies that use them for inappropriate decisions. To be able to negotiate more effective teaching the teachers and students must develop a culture of evaluation by a communal practice of *Situations of evaluation* in the course of teaching. They are the times for the teacher and the students to take stock, to look together over what has been done, what that means and what it would be best to do next. They are the instrument for transmitting a very necessary epistemology and scholarly *didactique*. We cannot describe how the teacher conducts this type of Situation until after we examine the conduct of a-didactical Situations.

The Types of A-didactical Situation and How They Are Conducted

The objective of a-didactical Situations is to induce manifestations of *connaissances* such as decisions (if possible adequate ones), formulations (effective whether or not correct), and/or convincing proofs characteristic of the notion to be taught. They take place before the phases of exposition of *savoir*. That way the *savoirs* become a conclusion that the students can draw, after some preliminary work that bears more resemblance to motivated research than to free exploration of a theme. This approach thus precedes (but note that it does not exclude) the classical presentation that proceeds from the study of a text to be learned and known (definitions, fundamental theorems…) to formal teaching, then to its applications.

The goal of these a-didactical Situations is to facilitate the learning of the corresponding *savoirs* by first making familiar and intelligible what it is that they mean, which is what the students ultimately need to acquire as canonical knowledge. The formal classical learning comes in as a supplement, after phases of intense use of the *connaissances*, motivated by other projects. This type of Situation must be distinguished from classical "discovery situations" in which the teacher has the students visit various aspects of a notion borrowed from a text that is already there.

The teacher must concentrate her efforts and those of her students on the questions posed and the tasks to be carried out and thus avoid creating a direct didactical tension about the *means* of accomplishing these tasks (the *savoir*). Learning is a spontaneous consequence of the activity. The *connaissances* are thus the means and not the official goal of the Situations. At the same time, they are also one of their consequences.

Once a *connaissance* becomes sufficiently familiar it is time to recognize its importance and its place. The students then may well be willing to make an extra effort in the form of exercises "to make it stick" in order to make their use of this recognized *savoir* easier and more fluid. Learning snippets of knowledge under construction head-on and without relevance in order to apply them in conditions as yet undiscovered – the classical method – has its points, but it requires of the students a great deal more confidence, attention and good will.

The choice to teach the use of the *connaissances* before making them the objects of *savoir* to be learned voluntarily is a positive one. It relieves the students of the tension created by the obligation to regard everything presented at any moment in class as equally indispensible and decisive and hence in need of instant learning. Each lesson is thus the occasion to make progress with some *connaissances* and among them to recognize some and institutionalize them, and to exercise others that have already been institutionalized.

This choice requires of the teacher both sophisticated methods of evaluation and complex decision strategies.

Situations of Validation (or Proof)

These are the ones that make the mathematical reasoning of the students most visible, as they produce arguments addressed to their peers with the goal of convincing them. It's a matter of inciting the students to become skeptical about some precise mathematical notion and of giving them a motive and the means not just to check the validity but beyond that to convince the other students.

These Situations develop the capacity to produce, appreciate and judge arguments and in the end to distinguish and reject incorrect rhetorical methods and practices. In organizing debates, the teacher also teaches progressively more formal rules. On the other hand, it is essential not to lose sight of the fact that the important thing is the declaration and its proof. This type of initiation rests principally on the cleverness of the teacher, whose interventions must be attuned to a variety of indices in order to optimize the interest and participation of all the students. This cannot be judged solely on the number of participants, nor on the speed with which the solution is given and established.

She can for instance, organize debates first in very small groups and then in larger groups to bring up alternatives. Whatever the format, the game must be worth the time and effort required for it. The interval between too obvious and too complicated can be a very narrow one. If the whole argument depends on an abstract demonstration, the discussion may degenerate into a debate among two or three "champions" without benefit to the most of the class. Speeding up or slowing down the process, maintaining the engagement and pleasure of each student, avoiding traps posed by individuals, cutting it short or waiting patiently – only a report of the discussions of the debriefings of the teachers and the researchers could do justice to the subtlety of conducting this kind of lesson, often halfway between reality and simulation, and to their influence on the enthusiasm of the students when they were successful.

Situations of Formulation

In order for a Situation of validation to function, the students need to have understood the object of the debate and thus to be capable of formulating the elements of it. Some specific Situations lead to this result by challenging the students to communicate some real information to a partner, either using a known vocabulary or by creating a repertoire or even a provisional specific language like the one for designating the thickness of the sheets of paper or of the enlargements. This type of Situation where actual communication is organized requires particular set-ups and materials for the class and thus must not be trivialized. But conducting them is much less of a burden on the teacher. The times and results are easier for the teacher to regulate and the students to evaluate. Furthermore the students rapidly acquire sufficient knowledge about communication for the teacher to be able simply mention it without actually realizing it, and save actual communications for the cases that merit it.

Situations of Action

Formulations only make sense through their meaning in terms of decisions in a specific conditions. Thus the whole construction is founded on the possibility of giving each mathematical notion to be taught a meaning that is simultaneously significant, correct and fruitful. This meaning is traditionally given by the linguistic means offered by the culture: verbal definitions, explanations and proofs – essentially by texts. The teaching of mathematics is thus reduced to the study of a text with the aid of texts that may be illustrated by a discourse. These means appear economical because they facilitate the communication of the text of the *savoir*. But in reality they are not economical for the students, who grasp concepts better by their function in the course of an action in a situation and by the decisions that it calls for than by descriptions and intellectual proofs. Action Situations, in the large sense, are thus the foundation of the whole edifice for all of the students.

Carrying out an action Situation was fairly easy because it had been well conceived. The teacher had to restrict herself to being satisfied with the first successes. She was not supposed to approve them or spread them around. With the complicity of the students who had found an answer (which they thought was right, or knew it was because it obtained the desired result) she encouraged each of the students to try to find it. And she received them all equally, whether they had been invented or inspired by an auxiliary peek at a fellow student's work. The essential thing was that the student adopted a production as his own. Who remembers how and from whom he learned the words and most of the knowledge that he uses?

The principal difficulty for the teacher in conducting an a-didactical lesson is maintaining a fragile equilibrium between what is said and not said, what is desired and what assumed, what is suggested and what required.

For the students, the *connaissances* thus emerge from a story resulting from a mixture of truth and fiction. The story told at the end of the adventure by the students and by the teacher assembles these pieces and becomes not just the reality of a class but the legend of the birth of a notion or a concept. The important thing is that that adventure be intriguing and fascinating, that it be possible to engage in it with one's strengths and weaknesses, and above all that it have a meaning and an epistemological and didactical value such that the quality of what is gained justifies and recompenses the efforts, the disappointed hopes, and the vain attempts.

And what happens when for one reason or another the miracle does not happen? Sometimes because of a detail, a fault in the preparation or execution of a delicate sequence of actions produces a fiasco: the materials refuse to follow what appears to be their natural law, damp paper compresses and eliminates all precision of measurement, the water spills (predictably!) over the side of the bowl, the pantograph does not work right or a bird flies into the classroom … everything gets muddled and nobody understands anything, or worse understands the reverse of what was hoped for.

Nothing is lost and often the students not only imagine and understand anyway what was supposed to happen, but sometimes even understand it better than if they had gotten it without the complications. And depending on her personality, the teacher repairs the thread of the story they are in the midst of in her own way, and admits that, like her students, she cannot always get everything right.

By frequently putting the teacher and the students under the obligation of cooperating to make the current action succeed, Situations stimulate, facilitate and guarantee a large part of the learning of the goal knowledge.

Situations of action, formulation and proof (or validation) proceed in principle without the teacher intervening directly in the course of their solution. They are called a-didactical: in them the teacher is not directly teaching any knowledge. But they should most often be proposed by the teacher, who ought at least to "teach" the rules of the game as instructions – the students should simply learn to play, not take the rules as *savoir* to be learned. The teacher *informs* the students and *prescribes* an activity for them. (He introduces the rules to be followed and an objective to aim for as a provisional institution – a convention – in the class.)

At other moments the teacher may intervene to *comment* on the progress of the lesson and to *report* with the students the state of the adventure and its results. Recognizing, organizing, presenting, explaining and leading an evaluation of the *savoirs* aimed at, drawing conclusions from these reports in terms of decisions for following lessons are types of didactical Situations (because what is taught passes through the formulation of the didactical will of the teacher).

Presentation of the Rules of the Game

The teacher transmits the rules of the game, but these rules are means of learning, not *savoirs* to learn. They may be forgotten, but in fact they leave a trace in the form of the conditions of the final *savoir*. The teacher proposes the Situations, which are in charge of advancing the class knowledge. She must present the materials, designate the players (individuals or teams), indicate the goal of the action of the students, the starting position, the activities permitted or not permitted, the final state being sought for and the states that indicate a failure. She might offer a reward – a purely symbolic one – or designate the number of rounds to be played.

If the students are to undertake an action that might bring them some additional information they must envisage a *basic strategy*, which the teacher might possibly suggest. In general, this basic strategy is not the one that is supposed to be found.

It won't work fully, or it is long and messy, and it should swiftly be clear that it needs to be avoided. It simply makes it possible to take the first steps.[4] Note that one must accept as a success – as mathematicians do – the blind trial of all possible cases, or even the presentation of a good solution when the student is unable to explain how he got it. It is enough to show that the solution is valid and sufficient. The rest (commentaries, explanations, etc.) is a legitimate requirement, but it is didactical.

Except for exercises and classical problems, the teachers almost never simply set out the instructions for Situations in a form that had been written up for them. They needed, for example, to play a couple of trial rounds so the students could understand the rules. The ratios between the time spent explaining the rules, the time the students needed to solve it, and the importance of the knowledge that they needed to use to do so were clearly decisive criteria.

The teachers must above all pay attention to the time required for a Situation. If the "Situation" under discussion is such that the students could find the strategy and answer without actually playing a round, then it is just a question and should be treated as such. It is better not to use a game if:

- The rules are harder to teach and understand than the solution
- The solution cannot be found in a reasonable time (then it is just a riddle), or
- It does not require that the student invent an interesting and instructive strategy (then it is just a pastime).

Evaluation in A-didactical Situations

A-didactical situations mobilize knowledge that the students are in the process of learning. Thus they constitute an opportunity for the teacher to evaluate the acquisition of that knowledge. But this evaluation is not summative, it is formative. The student carries on without thinking about it if he succeeds. If not, he simply notices that things are not working and either fixes it himself or calls on the teacher, who can record the fact, but with no immediate consequences. The teacher also goes on some indications that the student is unaware of: what the latter does and says must be interpreted. The student's knowledge evolves differently from that of students in situations where the learning is parceled out and the evaluations match the parcels. Nadine Brousseau's excellent descriptions that keep us in contact with the students in the class in Chap. 2 could not include the mass of individual and collective observations that she collected and decoded instantaneously to understand the state of the Situation and evaluate its consequences in order to decide whether to intervene immediately, or delay intervening, or not intervene at all.

---

[4]This method is comparable to the attempts to prove Fermat's conjecture before the twentieth century. Working with a fixed, particular value of n was clearly not going to advance the general solution at all, but there was always the hope that the examples would provide some useful reflections.

The precision, quality and dependability of the Situations made it possible to reduce evaluation to observation of how they actually developed and the participation of each student. This development was reproduced regularly each year with the same results, which contributed greatly to reducing the fears of both the teachers and the students to an acceptable level.

Obsolescence

Situations, whatever they are, are adventures for the students each new year, but for the teachers they tend to become rituals. The teachers' memories lead them to construct a simplified and stereotyped image of the development of the Situation.[5] They are then unable to respond in a differentiated and opportune way to the actual events that occur, or even to let them occur. The Situation becomes a classical lesson that then loses its suggestive properties and thus becomes far too heavy for a minimal profit.

In trying unconsciously to have the students reproduce a stereotyped development, the teacher tries to prevent the difficulties observed in the years before. She intervenes more and more directly in the behavior of the students and the Situation becomes purely didactical. She tends to transform high level objects (for example those corresponding to high levels in Bloom's taxonomy) into algorithms. For the student, the situation loses its suggestive qualities and becomes the execution of a sequence of instructions, a simple task.

The relatively unpredictable character of a-didactical Situations helps the teacher fight this tendency. Nonetheless, Nadine Brousseau notes that she had to make an effort to maintain her capacity to deal with diverse but equivalent manifestations of the same knowledge. For example, she feels that it was in the end a good thing that she had to accept and monitor reasoning about fractions and also commensurations, which were fairly unfamiliar to her. It is important to distinguish between what is justified for the students and what is obvious to the teacher.

The complexity of the evaluation, interpretation and management of the a-didactical phases of the acquisition of knowledge may explain the evolution of the practices towards exclusively didactical methods as the pressure of evaluations mounts. A heavy tendency has been observed since the 1970s to replace the phases of acquisition of *connaissances* – which by definition should always precede the teaching of *savoirs* and the evaluation of the ensemble (which had developed greatly in the previous century under the influence of great pedagogues) – by direct instruction of answers to questions on standardized tests. Nicely aligned with naive popular beliefs, this practice greatly simplifies didactical decisions (start over, increase the pressure, eliminate comprehension in favor of reproduction, discriminate among the students, individualize, formalize, etc.), the knowledge needed to make those decisions and their justification with the population. But no observable improvement in results of teaching has resulted. As standard evaluation becomes more and

---

[5]This phenomenon is being studied under the name of "Obsolescence of Situations."

more present, with higher and higher stakes, it tends to produce the progressive disappearance of all the activities that traditionally preceded or accompanied the construction of knowledge.

Even though they cannot be evaluated in a standard way, *connaissances* are nonetheless indispensible. They should always accompany the teaching of *savoirs* and their evaluation. Training for standardized tests by giving standardized tests (teaching by worksheets) destroys the coherence both of the mathematics and of the class.

Isolated Evaluation of *Savoirs* and Constant Evaluation; the Necessity of the Uncertain and the Implicit

The evaluation of *savoirs*, the kind whose outcome is known to both the teacher and the students, is only justified in the case where one or the other of the protagonists has available not just the means of judging the results but also the possibility of using this outcome as a basis for worthwhile decisions. Otherwise it is simply a question of unjustified and unhealthy pressure, by definition unproductive.

The evaluation that plays an essential role in the playing out of the lessons is the one that the teacher and students engage in separately. It takes the ambiguous form of assessments, encouragements, questions, funny faces, etc., using a complex and delicate system of communication.

The Play of the Real and the Fictional

The curriculum gives the teachers the canvas and the means to make present a story that constitutes a sort of epistemology of each notion. But if they apply the program of work without discernment they will devote a considerable amount of time to episodes that are of no interest to the students and/or not productive of much learning. One means of regulation at their disposition is the passage from real mode to fictional mode and vice versa. Students can understand a lot of information without its having to be imbedded in an actual action on their part and a fortiori in a situation that can be complex and difficult to put in action. When a Situation that demands a certain intellectual and material investment has its effect, the students suddenly understand the notion sought for. They imagine the possibilities of what follows, anticipate the didactical intentions of the teacher, and often negotiate with the teacher to abandon the action phase. Their relationship with the Situation becomes imaginary. The imaginary mode makes it possible to save a lot of time – but if the students have made a mistake there is none of the feedback that a real Situation would have given them. Explanations may then be long, tangled, delicate and hazardous. Actually checking things out physically frequently meets with resistance from the students; re-checking an idea in a real Situation meets with even stronger resistance. The students ask the teacher to hand them the answer. That is why real, costly Situations must be rare, fascinating and productive of emotions, questions, etc.

For that, the teacher absolutely must retain control over the passage from one to the other: the real mode is slower but surer and the imaginary mode is faster and more productive. They are alternately indispensible. The teacher must maintain an optimal balance between a reasonable speed and a reasonable comprehension by the set of students.

In this dilemma the teacher needs the consent of all of the students. Those who have found a solution or think they have should wait until the others have had a reasonable time to carry out their own actions; those who are having trouble should feel pushed to do it right by the fact that the others are waiting. The acquisition of knowledge is a collective effort, like a culture. Individualism makes the work of the teacher and the students very difficult. Emulation stimulates but does not help, while cooperation helps, especially when the work of some depends on the work of others.

The Inexpressible, the Said and the Unsaid

The conduct of Situations is much more sensitive than classical lessons to the maintenance of a suitable equilibrium between what is or should be said and what is not, or should not be. Knowledge and invention or learning are the means of reducing the uncertainty presented by the Situations. The teacher must at all times monitor the part of the help that he offers the student to help her advance and the part that he leaves for her. In this paradoxical relationship where the teacher must say everything about what he wants to teach except for the most important thing: what he wants to teach the student to do and think herself, the unsaid and complicity play a considerable role. Any attempt to clarify everything and require immediate describable and measurable results like those in a notary's contract immediately condemns the teaching and learning project to irremediable failure. Furthermore it is abusive. A-didactical Situations let the teacher stand beside the student and follow her efforts, leaving to the Situation itself the task of criticizing them. And students accord far more importance to things they have flushed out themselves than to things procured for them without any effort on their part.

Further Aspects of the Teachers' Adventures

This chapter presents only the didactical part of the adventure of the teachers engaged in this COREM project. Other aspects would also have been of interest, among them the work of the teacher: the preparation of ordinary lessons, that of "experimental" lessons, lesson development as a team, difficult lessons, evaluations, commentaries at the moment, seminars, mathematical training, relations with other disciplines, their point of view about the students, and also especially their relations with the students and their parents, and the authorities, also their relations with the other teachers.

Their roles as teachers and educators were complicated terribly by the requirements of the research: having always in "their" classroom two colleagues with whom they had to share the attention, the obedience and the affection of the children; the obligation to coordinate with and thus to explain to the colleagues the intentions and results of the lessons; the intrusive attention of the researchers, the hide-and-seek game with what needed to be understood of their intentions, their implicit or explicit requests and what they could not say because knowing that a result is anticipated makes it almost impossible to avoid intentionally or unconsciously working towards it. The subtlety of these relationships created and was sustained by intellectual complicity.

The teachers had the responsibility of defending the interest of their students. They had the last word whenever that was an issue. Researchers and teachers strove to conceive at all costs situations that would teach what was necessary for the students, but the designs were such that the scientific conclusions were never at any time based on failures of the students.

Each experimental lesson was an adventure for the whole community, students, teachers and researchers, and there was no need to add to the hazards supplementary recompense or a fortiori sanctions. The quantity of observations on the complex set of scientific questions in course was abundant for everyone. The community appropriated them and kept them available for the next adventures. So-called "ordinary" lessons might equally well harvest the most obvious and certain conclusions of the current research or be directly and narrowly inspired by the most down-at-the-heels pedagogical or didactical models. What was called a "basic lesson" was composed in the most classic manner of: mathematical terms, presentation, text, examples, exercises, explanations, problem, applications. But starting early on the teams of teachers often had to modify them to benefit from the opportunities created by the experimental lessons. They had that liberty, and felt free to exercise it on the basis of lines that they found solid and practical. And the researchers in their turn exercised a certain vigilance. But not one ideological or systematic slippage, not one hasty generalization based on one or two "successes", not one general rule had the right to be indulged. Filmed observations are a cruel threat to that kind of slippage.

Also of interest is how the teachers accustomed themselves to a new vocabulary for concepts that they already knew, and for new concepts sometimes behind the same vocabulary, and also how it was necessary to fight against "didactical permeability", that is, the uncontrolled and regrettable penetration of scientific didactical (or psychological, or other) vocabulary into the exchanges of the teachers with their students.

And also, how did one get into this establishment? Teachers, researchers — who was chosen, or rejected, and why? How were things worked out? How could the administration accept such a singular teaching environment? Did it cause any difficulties? These adventures are part of the same story, but they would need another book!

In any case, to this day no one has done justice to these teachers for the treasures that they inventoried and put at our disposal. The adventures that this book describe were hose of all the members of a unique institution. In the course of 25 years it involved more than 250 adults and nearly 2000 students.

The Mathematical Organization of the Curriculum

We have now looked at the curriculum from the perspective of the students and the teachers. One important vantage point that remains is that of the mathematical foundations on which the whole sequence is constructed. The mathematical plan underlying the curriculum is that of a rigorously mathematical construction of the positive rational numbers in the modern sense: it is axiomatic and based on formal mathematical structures. But this plan is also subject to a complementary set of conditions of epistemological and didactical origin. Thus in *Part One* (the first three modules) the positive rational numbers are introduced as a set of numbers designed to measure lengths, masses and volumes using an arbitrary unit, and to provide, by a calculation, the results of the physical operations of addition, subtraction, and multiplication, and of division by a whole number.

In the *second part* (Modules 4–7) the children become conscious of the difficulties of effectively putting fractions in order, of estimating their differences, of locating them and of keeping up the habits developed in dealing with the natural numbers for all the customary operations of measurement. And it is they themselves who choose what mathematicians call the *decimal number filter* to "represent" the rational numbers, or more accurately, to approximate them with a manageable precision. The game they play gives this search the meaning of "finding a decimal number as close as is necessary to represent a given fraction." They end up having available the remaining necessary operation – successive divisions of natural numbers – as a unique operation that looks like division but will only be recognized as such after some other adventures. The repetition of the operations and of the reasoning about calculations gives the students, even the less swift ones, a chance to carry out a number of useful and instructive operations.

Classical curricula treat the decimal environment as an obvious extension of the practices of natural measurement and are content here to teach algorithms without mathematical content, as simple conventions. In these curricula the mathematics appears after the fact, simply as a commentary, or else as a refinement that breaks with the previously inculcated practices. In those conditions it can only strike most students as casual remarks of no particular interest.

The *third part* (Modules 8–11) introduces rational numbers as functions and as scalar ratios. The lesson on the Puzzle makes this introduction the object of a new adventure on the conditions for the conservation of ratios. They thus define linearity by a non-mysterious criterion: the image of a sum needs to be the sum of the images, in contrast to the traditional reference to proportionality, which is more mysterious and always gives some students trouble. The study of geometrical forms swiftly provides the occasion for extending the practices they have been using with rational and decimal numbers as measurements to a set of functions. Working on putting the enlargements in order makes them revisit rational and decimal numbers and think about changes of unit and reciprocal mappings. The structure thus constructed is that of fractions as scalar operators.

Next, in the *fourth part* (modules 12 and 13) the search for new uses for linear mappings leads the students to rediscover the everyday uses of fractions (percentages, scales, taxes, etc.) as well as the translation of operations and the raft of specialized vocabularies associated with them (for example "taking a fraction of something" as a way of saying to multiply by it.) This swiftly leads to the study of external linear mappings, that is, mappings between quantities that are of different natures, and hence accompanied by a dimensional equation. Some of them the students already know well (price/quantity), others are new (distance/fuel consumption, speed, density, debt, etc.). We include the classic use of these visits in their role of review, of illustration and enrichment of concepts, of learning exercises, and of initiation into the ordinary use of elementary mathematics.

The competition of problems posed by the students gives them a chance to pose problems and discuss what is interesting about them (and not just to answer them). We try to develop their interest in problems and the culture of problem-setting. This is an occasion for revisiting all the interpretations of division.

The *fifth part* leads the students to consider, use and calculate compositions of linear mappings, their decomposition into natural mappings, and their inverse mappings (they had already encountered reciprocals). They can thus express all of the interpretations of rational numbers – as measurements, ratios and linear mappings – with the same symbols, those of fractions.

A *sixth part* was prepared, but it was never possible to experiment with it, because it would have had to take place in the first or second year of middle school. It consisted first of symmetrizing the additive group by creating negative rational numbers and completing the construction of the field of rational numbers. The introduction of algebraic symbols then made it possible to formalize the definition of certain useful meta-mathematical terms and of the proofs produced spontaneously in primary school.

Mathematical Commentary on Chap. 2

Even reduced to its mathematical structure the curriculum presents numerous points that may appear strange and even difficult to accept, both for teachers and for mathematicians. The former may be suspicious of the mathematical quality of notions brought up so differently from the normal presentation, and the latter may contest them in the name of the didactical culture they remember from their childhood. For example, it is well known that the use in primary school of algebraic notation like $3+4=7$ does not give students the meaning of an equality. It has been demonstrated that this practice develops in the children an erroneous comprehension and use of the $=$ sign that perturbs the mathematical practices of students all the way up to the university (where they are seen to understand and prove the same equation differently depending on the order in which the two members of it are written.) But an attempt to replace this notation with $3+4 \to 7$ would give rise to great protests from different people for different reasons.

Elementary mathematics is frozen into practices that we are in the habit of regarding as untouchable.

Here is a description in scholarly mathematical language of the activity proposed in Module 15 of the curriculum to these 10-year-old students: they have equipped the set of rational linear mappings with their multiplicative, commutative group structure, distributive over the additive semi-group of natural numbers. It is a jolt to see it written this way, because it gives the impression that the students are going to have to learn this vocabulary. This is not the case. But it makes it possible for the teacher to verify whether among the exercises proposed all of these properties have indeed been used and justified – which does not in the least necessitate any meta-language beyond ordinary language or any other explicit proof than the comprehension of what one has done. This gives a legitimacy to ulterior definitions (the way language justifies the study of grammar).

Whether or not to teach meta-mathematical terms to students is a much debated question.

We have picked out a few of the singularities that solid mathematical and/or didactical reasons have led us to prefer to more classical practices, and we will give an elementary mathematical justification for the teachers that the mathematicians can easily verify.

The Temporary Replacement of Fractions by Commensurations

A stack of T identical sheets of paper has a (whole number) thickness E. (T,E) is an ordered pair. The students first posit that two ordered pairs (T,E) and (T',E'), each consisting of identical sheets of paper, both consist of sheets of the same thickness of paper if there exists a number a such that $T'=aT$ and $E'=aE$. This condition is sufficient but not necessary. A necessary and sufficient condition is that there exist a and b such that $(aT, aE) = (bT', bE')$. $(aT, aE) = (bT', bE')$ is equivalent to $aTbE' = aEbT'$, and thus to $TE' = T'E$. This will be used explicitly and even proved (without algebra) by the students when they get to commensuration of lengths (module 3).

The ordered pair (7,4) indicates that a stack of seven sheets has a total thickness of 4 mm. The thickness of a sheet is expressed by the commensuration 4/7 or by the fraction 4/7. The words commensuration and fraction are synonyms and share the same symbolic notation. The conceptions are not.

Commensurations use only operations that can be conceived, realized and carried out materially or by calculations in the known domain of the natural numbers. Fractions make it possible to go back to the familiar model of the natural numbers by using an intermediate unit, but one must assume the prior existence of unit fractions, that is, fractions of the form 1/n, which cannot always be easily constructed. Historically, fractions are the concept that has been retained in all cultures.

The concept of commensuration could not arise from the manipulation of concrete lengths long enough to permit the use of a sub-unit obtained by folding and repetitions. Hence our use of a unit that is essentially indivisible for children: the millimeter, and of a length for them to measure that is smaller than the unit. This artifice having rendered the use of fractions improbable, the students were able to invent an original solution to a completely concrete problem and thus undertake the exploration of a necessary mathematical knowledge before being able to name it, justify it and recognize it as a familiar concept.

2. Introducing the topology of decimal and rational numbers as we do in the second part of the curriculum looks as if it were an ambitious and useless enterprise.

Tradition offers students an amalgam of various vague structures mingled under the name of "number". We will point out four.

- Natural numbers. They are more or less correctly understood, but the disappearance of analogical instruments of measurement has caused the disappearance of a powerful means of teaching the topology of the natural numbers which served as a basis for that of the decimal and rational numbers.
- The algebraic structure of the positive rational numbers is taught, even though they have been profoundly scarred as a result of hesitations and accidents in their history. Students distinguish them easily from natural numbers because they are written in the form of "fractions", and calculations with fractions are studied. On the other hand, comparisons, the concept of intervals, and complete ordering are ignored. Thus their dense topology, the first simple reason for their existence because they provide a means of attributing a distinct value for every distinct measurement, is neither practiced nor even envisioned.
- To top off this gap, students use the notation of positive decimal numbers, but the way they implicitly conceive of the set of these numbers leads them to make errors in their subsequent mathematical studies. They reason as if there existed a unique natural number n such that if all decimal numbers were multiplied by 10^n they would all become whole numbers (Such a structure is called D_n.) Topologically, thus, they are still just the whole numbers.
- The real decimal numbers are all the rational numbers of the form $m/10^n$, and thus the set of all of them is the union of all the D_n. They can approximate as closely as desired not only any rational number, but also any algebraic or transcendental real number. They lend themselves to algebraic calculations like the rational numbers and to comparisons and ordering like the natural numbers.
- Students and sometimes even teachers use the term decimal number to designate any *decimal expression* of a rational number, be it a decimal rational number like 0.3 (3/10), or a non-decimal rational number like 0.333... (1/3) or even the decimal approximation to an irrational real number like 3.14 for π.

• These mathematical delusions do not prevent the students from being disconcerted when suddenly the division of two well-behaved natural numbers sets out to produce a sort of monstrous number, made up of a visibly infinite sequence of digits.

Part 2 of our curriculum therefore shows the students the coherent project of replacing the antique fractions, inappropriate for analysis and calculation, with decimal numbers that can easily generate the real numbers. The properties of one and another can thus be established by an authentic and instructive mathematical adventure that they will perhaps recognize later on in more sophisticated guise if they study mathematics.

The students compare and calculate a large number of decimal numbers and intervals, imbedded or not, and rapidly develop an expert *connaissance* of the real line, a *connaissance* that the exclusively numerical apparatuses in their environment no longer show naturally.

In Part 3, the three fundamental objects represented in the course of the story by fractions or by decimal numbers remain distinct: *measurement of sizes, ratios,* and *functions.* The operations on these mathematical objects are conceived differently, and the names for each, depending on their particular uses, proliferate. The problem of change of the unit in the reproductions of the Optimist poses an apparently difficult problem for the students. Clearly, as before, the procedure that they will be using and that will work is not taught to them. The students know very well that they will not have to reproduce them alone and in other conditions. What they are doing has a mathematical identity that can be expressed in more advanced terms in later mathematical programs. The issue is not anticipating these advanced *savoirs*, but justifying the use of the necessary instruments that one wants them to learn by a mathematical problem that gives these notions their meaning and their mathematical use.

In Part 4, using the composition of functions in order finally to define multiplication and division of two fractions appears a new challenge to caution and reason. This new case of division gives rise to a cohort, nonetheless already numerous, of different interpretations. In all of the curriculum, divisions are the quick-change artists of this adventure: they keep reappearing in new guises, for apparently similar (or inverse) uses. But the final identification of all these appearances gives the students a very satisfying sense of accomplishment. A cycle of study is achieved, producing a sentiment of simultaneous completion and unity.

Chapter 4
The Adventure as Experienced by the Researchers

This chapter, originally conceived as a discussion of the research intentions behind the lessons in Chap. 2, how the research was carried out, and what results were developed, underwent a major expansion and enrichment. These resulted from the fact that in going back to produce this discussion, Brousseau became conscious of a degree of complexity that was more apparent in retrospect than in the process of living it. Delving into that complexity broadened and deepened the discussion.

Warfield Introduction Concerning the History of and Voice in Chapter 4

Our original image for this book had three parallel chapters, all centered around the curriculum on rational and decimal numbers that was at the core of the Theory of Situations, and a foundation stone for the whole research field of *Didactique*. The idea was to start from a students' eye view with the lessons themselves and some of the commentary written at the time about students' responses, augmented by Nadine Brousseau's memories from having taught them. In this we were aided by the fact that we had published many of the lessons and a lot of the commentary in a sequence of articles in the Journal of Mathematical Behavior. The following chapter was to be the teachers' eye view of the curriculum and the process of preparing and teaching it. This proved to be slightly more complex than we had anticipated, because the setting was so important and so totally different from a "regular" school – French or not – that it needed a lot of explaining. Again, the Brousseau's put their heads together, and eventually developed a chapter that covered the basics in a clear and honest way.

Then came the final chapter, the researchers' eye view. The writing of it has proved to be a major and also highly rewarding challenge. The challenge stemmed basically from the fact that as he began to describe the researchers' view, Brousseau realized that in fact he was describing a piece of the history of a research field which

G. Brousseau et al., *Teaching Fractions through Situations: A Fundamental Experiment*, 165
DOI 10.1007/978-94-007-2715-1_4, © Springer Science+Business Media B.V. 2014

had never been published and which only he actually knew. From this realization grew a determination, strongly supported by his co-authors, to get this history down on paper and make it available. Doing so proved to involve an enormous amount of work and sometimes frustration. Repeatedly he read what he had written and, feeling that some major aspect had been omitted, in effect started over. This process lasted for many months until finally he felt that all of the essentials had worked their way in. I then had the privilege (and task) of assembling the results of all this labor into a coherent whole. In this I was hampered by the fact that I was unwilling to eliminate any pieces, because each piece provided at least one insight that was fascinating and quite new to me despite 20 years of working closely with Brousseau and his work. My hope is that readers will be as fascinated as I, and correspondingly excuse some slightly awkward transitions.

As should be clear, the chapter also escaped our earlier intention to avoid the first person singular. It is the story of the personal efforts of one person, strongly supported by a multitude of others. Those efforts led to the establishment of a whole field of mathematics education research, whose value was ultimately attested to by Brousseau's reception of ICMI's first Felix Klein Award for Lifetime Contributions to Mathematics Education. The rest of us have the privilege of hearing the story directly from that person – it would make no sense at all to attempt to disguise that fact. The rest of this chapter is firmly in the voice of Guy Brousseau himself.

Brousseau Introduction to Chapter 4

After describing the adventure of the students and then that of the teachers, I thought it would be easy to present the principal results of our research as an adventure of the researchers. This turned out to be impossible. Too many observations took part at the same time on too many concepts engaged together in too many concomitant experiments. To describe and explain each step would require the description and explanation at that moment of all the other concepts in evolution and the effect of each event on all these concepts, and why progress on one point sometimes disrupted all the rest.

All of my efforts to explain at a distance of 50 years the emergence of the principal concepts that today make up the Theory of Situations have failed. Each attempt led to some new sort of exposition, much more detailed than the preceding, of the set of concepts created, constructed according to the normal rules of scientific exposition: definitions, statement of results, proofs, consequences, uses, … I never stop reorganizing and completing "a complete exposition" of the Theory of Situations and its extensions and metamorphoses. But invariably this exposition presents gaping holes resulting from uncompleted research projects, concepts that outside of their original conditions have definitions that float, or incomplete proofs, …

As a reader accustomed to scientific reports I thought that my difficulties arose essentially from the confusion of my ideas, increasing with age. But that is perhaps not the only explanation. As I revisited certain of the episodes from this adventure on the teaching of rational and decimal numbers I suddenly rediscovered a difficulty that I had experienced vividly without ever having brought it to the surface and straightened it out.

The objects of the Theory of Situations have a composite origin, and they are held in "logics" that are locally irreconcilable. There is what one wants to do and what one sees or thinks to have seen. For example a *didactical situation* is conceived as a means of realizing a certain didactical project. In this case, it is organized as a function of an end goal; the connections between the designs are themselves the reasons for the designs; the objects presented are assumed to be observable and realizable; the causalities are assumed to be true in the name of a previously established rationality. But the observations of an episode[1] must be interpreted. The elements of the study must be extracted, based on the observation, *independently of the initial model,* of the teacher, the students, the knowledge, the actions, the words spoken, … The gaps between the actuality and the elements of the model will only be noted later. Reasons can no longer be invoked but must be proved. Using the same terms in the two cases produces the risk of confusing the two objects. For the people carrying out the teaching and the research confusion was impossible. It was prohibited by the context. The theoretical concepts of situation, of a teaching project, and of observation of a phase of a lesson, even though spoken about with the same terms were necessarily different objects. But when they were discussed, these distinctions disappeared without our noticing it, and made the discourse difficult to decode.

With concepts like that of the didactical contract, the confusion was between the occurrences observed and the interpretations of them, and their technical and theoretical models. For some concepts, people were reduced to total confusion because of this multiplicity of statuses.

Scientific practice would have one consider the models as direct representations of observation, with the researcher charged with rejecting the model in the case of a recognized divergence. But that image is simplistic. Each science has the obligation of insuring the basis of the relationship of its discourses with a certain set of circumstances. The theory of situations was a method, a justification and a guide for the teachers and their counselors; for the researchers it was a hypothesis, a model that might be disproved and rejected.

The constant contact and blending with our object of study obligated us to take into account ethical considerations: not to use designs that risked producing unnecessary difficulties for the students; not to base our research on the interpretation of errors that we could avoid; to stay conscious of the effects on the system that would be produced by publication of the errors we observed; and finally to advance criticisms publicly only with not only a conception of their correction but also the real and practical possibility of avoiding them.

I finally gave in to the evidence after numerous attempts that invariably led me to revise each preceding explanation of the origins of the concepts of *Didactique* of mathematics, clarifying it or correcting it: my attempts at an exposition

---

[1] Alain Mercier judiciously proposed that the name "episode" be used for an object of observation that permits the hope of identifying a situation and its evolution.

suffered especially from a fundamental and insurmountable error, from a crucial misunderstanding:

At the time that the events we are reporting occurred, students and teachers formed perfectly identifiable entities. On the other hand, researchers didn't exist. The very idea that one could carry out specific scientific research on the teaching of a mathematical concept was inconceivable for the majority of scientists, first for the mathematicians themselves and also for elementary and secondary teachers. Journals that could publish the texts we produced did not exist, at whatever stage we presented them. The few rare documents that remain from the period are islands, peaks emerging from an undersea mountain range. There were no researchers in the sense in which we would understand them today. Describing the development of the science of *Didactique* as an adventure of researchers would have been a misrepresentation if it had been possible. It was not possible.

Our adventure took place at a time when *Didactique* did not exist, and no one knew what would constitute a "result" of *Didactique*. There were no journals that could publish the preparation of a "lesson", with its expected techniques and technologies, and at the same time a report on its observation accompanied by scientific references. The general ideas that supported didactical techniques and their observation were (or claimed to be) "new", but they would have had to be sufficiently sure, and thus already accepted, to make of these observations scientific results. In fact, they were radically different from classical professional concepts and from those imported from existing disciplines.

Every participant in the COREM nourished the communal knowledge by bringing in his or her observations, questions and problems. They learned thus what interested the others. No one had the total vision of what was happening, but each possessed a part of the truth that might call into question all the ideas, all the projects, and all the hypotheses. In the course of the weekly seminars all the participants studied or explained the works under way. But to act, each one had to be able to ignore, voluntarily or otherwise, things that might hinder their work. A teacher who waited too long, wanting to understand the behaviors of his students or perfect his explanations, might let the principal object of his action escape. A councilor or researcher who explained in too much detail the alternatives left open by her project took the risk of seeing the teacher derail during a lesson and take one of the options that had been studied and rejected.

The experiments were thus inscribed in the knowledge of the community without anyone possessing the whole of the necessary information, and the recognition of this knowledge was the beginning of putting them under study. The various types of difficulties or successes were thus experienced, recognized, then taken into consideration and finally studied in the collective functioning of the COREM.

The concepts, or their rough forms, had their roots in the work of the teachers. But the interpretations that the teachers made of them were oriented towards an indispensible rapid decision. And for that, these interpretations rested on principles, reasoning and observation that called forth objections and doubts. Not too many! Teachers could not be paralyzed by doubting their own judgment. Objections could thus appear only in the form of counter-propositions and the explanations they required. They had to be received and understood by everyone interested. The advantage of not having the status of researcher was that everyone had room for reflection, the rhetoric of reasoning, and experimental or at least empirical proof.

The researcher in *Didactique* in the years '64–'80 is a mythical being, a dream, a projection. I am the author, the inventor and one of the organizers of the system that I am attempting to describe, but each and every member of the system was also a researcher, a producer and an actor in this movement as much as I was. We were the idea-men, the authors, the organizers of the scene and the midwives for this mythical being: the researcher in *Didactique*.

Our model was Bourbaki.[2] Bourbaki was very probably, after Euclid, the most powerful "researcher" in didactical engineering of mathematics. He never established one new theorem; rather, he reorganized the known mathematics of his time. But his goal was to constitute a sort of new reference, a fundamental dictionary of mathematics unifying vocabularies in order to simplify its proofs, its learning, and its use.

He was also our foil. The virtues of this work enchanted those who had dealt with mathematics earlier. But newcomers saw classical mathematics suddenly becoming more distant – and for some even disappearing – behind a barrier of fundamentals destined, it was thought, to seat a future efficiency on a flawless rigor. Illusory efficiency and rigor. The diffusion of the foundations of mathematics seemed to put the encounter with its primary objects at an great distance. There were nonetheless many new ideas behind this apparent classical rigor. But they were not expressed. This engineering rested on an epistemology that was in the midst of evolving and not well incorporated.

Attempting to use the new axiomatic organization to reform the teaching of elementary mathematics could appear an extreme challenge and absurd gamble. But no! All of the conditions were indispensible to truly studying what could be the foundations for the ambition of teaching mathematics and discovering beyond the universal genetic development the causes, the effects and the laws of didactical efforts.

But if there were no researchers, who were the authors of the research? And how was it done? Driving the principal actor off the stage doesn't seem a good way to facilitate the presentation of a play… unless one considers, as is the case, that the story one is telling is primarily that of the ideas and events.

On the other hand, although the neat parallels between adventures of students, teachers, and researchers proved illusory, there is a good deal about the background of the research adventure that could contribute to an understanding of the adventure as a whole. In particular, it seems worthwhile to shine some light on the researchers themselves. What were their motives? What did they get out of the experience? What did they learn from it that was of interest to Didacticians and teachers?

---

[2] As defined by WIKIPEDIA: "**Nicolas Bourbaki** is the collective pseudonym under which a group of (mainly French) twentieth-century mathematicians wrote a series of books presenting an exposition of modern advanced mathematics, beginning in 1935. With the goal of founding all of mathematics on set theory, the group strove for rigour and generality." For more details, see http://planetmath.org/NicolasBourbaki.html

Our ambition is to render intelligible not just the results of our work, the ones that could be communicated in a directly accessible form, but also their genesis and their history, which were far richer than we have been able to communicate. This history will permit a better understanding of our results, their origin, their range, and the enthusiasm that enabled a small community to form and to furnish the efforts necessary for the realization of a project that was born at the end of the 1950s and continues today.

The first part of what follows shows how the main ideas for reforming the teaching of mathematics emerged: the search for modes of learning that permit students to recognize and develop the practice and knowledge of mathematical concepts before they can even formulate, define and explain them; the search for processes of reciprocal acculturation for mathematicians and teachers; the conception of organisms and designs that were capable of realizing the necessary research but that we did not have to develop into curricula that could be diffused.

The second part returns to the description of the COREM, this time from the point of view of the researchers: origins; objectives; types of research and methods of analysis; how and why this arrangement resulted in the appearance and testing of questions, concepts, phenomena, convictions and proofs related to a variety of aspects, in particular the mathematics of teaching.

Prelude (1960–1970)

The Sources

In the course of the nineteenth century the development of mathematics in all directions took a new turn: in a variety of areas the classical approaches ran into difficulties that called into question old practices and beliefs. The crisis of the foundations of mathematics was the most radical because it endangered the consistency of the whole structure. It led to an attempt early in the twentieth century to redefine and reorganize all of known mathematics. The hope that it would remain confined to the role of metamathematics, a description and redefinition of known mathematical objects, eventually died. The richness of this "language" and of the points of view attached to it led rapidly to its inclusion in a unified mathematics as the necessary expression of its objects. This then led to the question of how to give students of mathematics access to the mathematics of the period without losing too much time in the labyrinth of the old organization. At what level should the new foundations be taught? After the *licence*[3]? Then in every specialty – Linear Algebra, Topology,

---

[3] An academic level at the time very roughly equivalent to an American bachelors degree with a mathematics major.

Statistics,... – every professor would have to start with a course on fundamental structures! Absurd. In response, in 1956 the University of Paris inaugurated a course for first year students based on Pisot and Zamanski's book, *General Mathematics.*

In the autumn of 1957 I was called away from the classroom where I had been teaching to do my military service. While awaiting deployment to Algeria I was a sub-lieutenant at Fort Bicêtre, close to Paris, where my duties sometimes left me a bit of free time. I made use of it to enroll in the course to renew my contact with mathematics. Revisiting from a more advanced point of view the mathematics I had previously worked on was a pleasure for me. When I had to leave this mathematical interlude for less agreeable occupations I began reflecting on the possibilities for introducing traditional basic mathematics – the teaching of numbers and operations – which I had been practicing for 4 years, by basing it on this new definition. I began to understand the origin of some of Piaget's works (on which I had had to write a paper in the course of my training as an elementary school teacher.)

After 27 months of military service, 12 of them in Algeria, I rejoined my wife, my son whom I had barely met, and my class of 10-to-14-year-old students. I was able to carefully carry out the lesson projects I had been cogitating. The backbone of elementary applied arithmetic is proportionality exercised by cross-multiplication. Without mentioning the word "functions", I began to have them identify the values involved in a problem, and draw rectangles in which they could either write the given values or indicate the unknown values with a square (red for the value being sought.) Connecting lines indicated corresponding values. This correspondence frequently showed up as the possibility of carrying out an operation to find the number corresponding to a known number. Thus the notions of set and function were themselves introduced by their use, without the necessity of naming or identifying them for the moment. Manipulation of the numbers and of the units of measurement were designed to facilitate explanations and learning. I could only very partially touch on the idea of recognizing typical graphs for a problem, which could have introduced algebra, but it took me years to figure out why.

At that period, the ideas of Célestin Freinet were beginning to seduce teachers. They appeared to unite the propositions of the modern pedagogues by favoring student activity (Dewey), the grouping of activities around significant centers of interest (Decroly), respect for the liberty of the student and adaptation to the *milieu* (Vygotsky.) Nothing specific to mathematics was to appear. For example, Freinet proposes only that there be a folder where the student is to find statements of problems which he is to solve on his own before going to look at the solution. When he thinks he understands it he undertakes a test paper that the teacher corrects herself.

The Adventure of "Modern Mathematics"

At about this point (1960?) an article in the journal *Sciences et Avenir (Sciences and the Future)* revealed that in a course for future pre-school teachers Professor Papy had proved a theorem of advanced algebra (Bernstein's Theorem) using graphs.

This article brought to the attention of the public at large a movement known as "modern mathematics" whose goal was to revolutionize teaching. The journal cited the work of Lucienne Félix, "Modern Mathematics – Elementary Teaching". I sent Lucienne Félix my lesson plans for classes in which I had experimented with a number of her suggestions. She had me invited to a meeting of the CIEAEM (*Commission Internationale d'Études et d'Amélioration de l'Enseignement des Mathematiques*[4]) in Switzerland where I met the principal European promoters of renovation of the teaching of mathematics, successors to Choquet and Gattegno (among them Papy, Servais, Pescarini, E. Castelnuovo). Following this, I returned to my studies of mathematics at the University of Bordeaux and at the same time wrote a curriculum project for first grade. With no recourse to specific terms available, this pseudo-textbook was a 60 page book without words. It would have taken more than 80 pages of exposition to explain the mathematical notions taught or used, at least three times that in pages of methodology to describe to teachers what they should or should not do with it, and more than 1,000 pages to explain to them why they should and how they could try to change their practices without abandoning those that were indispensible. The publisher Dunot, at the instigation of the Academician Lichnerowicz to whom Lucienne Félix had introduced me, consented in 1964 to publish this emblematic, but practically unusable, work.

Professor Lichnerowicz directed my atypical course of university study. I filled it out with participation in the audacious activities of a young linguist, René La Borderie, director of the CRDP (*Centre Régionale de Documentation Pédagogique*)[5] of Bordeaux, who organized seminars with a parade of top-flight speakers (like Cristian Metz, specialist in the Semiology of Cinema, or A. Moles.) Thanks to this participation, and under the aegis of Professor Colmez, my Analysis professor, I was able thereafter to organize contacts and then professional development actions with the help of mathematics teachers from the region's Écoles Normales.

In 1964 I had made a lot of progress in my studies, but I still had a pretty heavy program in psychology with P. Gréco, and in Science of Education, and in Sociology,… and in every direction I looked. It was a Renaissance epoch, when all of Europe shook with exchanges and revelations in all domains.

It was at this point that Lichnerowicz suggested to me one day: "You ought to study the limiting conditions for an experiment in pedagogy of mathematics."

The Subject of the Studies Proposed by Lichnerowicz

I thought at first that he was asking for a sort of report and perhaps a catalogue of suggestions with a view to the reform movement which at that point was finally being widely accepted. That could have been a natural response on his part to my comments on the dangers and difficulties that I foresaw and that I talked to him

---

[4]International Commission for the Study and Improvement of the Teaching of Mathematics.

[5]Regional Center of Pedagogical Documentation. This center existed for the purpose of producing documents for use in teaching.

about, and to the solutions I envisaged. But the mathematical formulation of his request called rather for a deeper reflection and more precise justifications.

This request was the point of departure of the adventure of the didacticians that we want to cover in this chapter. But my questions were not of the type of "How many experimental and model classes should the administration set up, and what would be the budget for that?", but rather how to reconcile the flexibility necessary in order to adapt the project to a class with a respect for conventional conditions common to a whole cohort of schools – which notions were indispensible and how to make them accessible.

I had no intention of launching myself, like a visionary, into one of those exercises of Prospective so much in vogue at the time. Nor was there any question of immediately producing an elegant academic study, nor of enumerating theoretical conditions on which I had begun to reflect in which to set up my own attempts in classrooms. To discover the conditions of an experiment and experience the choices that would be required, the best thing was to prepare one and realize it.

But what *is* an experiment "in the pedagogy of mathematics"? It can only be the submission of an affirmation – expressed in a codified language subject to the requirements of a theory that guarantees its consistency – to an experimental process and to methods of questioning that could disprove it.

The conclusions that I arrived at fairly swiftly could have discouraged someone who expected short term results for immediate questions. But curiously I perceive now that I always acted as if while it was essential to work swiftly on avoiding predictable errors, it was necessary to take all the time needed to avoid going astray, and as if I had ahead of me … eternity, or in any case 45 years, to do it all.

I therefore launched myself with youthful enthusiasm in the conception and the progressive, meticulous and relentless construction of a potentially Herculean project. I had to begin to realize it in order for the circumstances to be able to impose on me the reductions to the essential: the initial conditions. Their realization would then permit the opening up of a process that would lead to the installation of institutions, the adequate training of personnel, and the elaboration of methods of study and knowledge necessary to the establishment of a science capable of framing a respectful reform of the object of its studies.

The work to which I progressively assigned myself was not that of a mathematician, nor that of an innovator, of a teacher or even of a researcher in the classical sense. It was more like that of a nineteenth century engineer who wanted to imagine and create an enterprise that he knew would be required in order to launch an apparatus that didn't yet exist – a flying machine, for instance.

The Background of the Future Research

Until the 1960s, the majority of mathematicians thought that the renovation of the vocabulary and fundamental concepts of mathematics could only be of interest at best to students who had already advanced well into mathematics. The most audacious wanted to extend the project to all students of mathematics at the universities and in particular to those who were beginning upper level studies. This state of affairs was rather abruptly reversed in 1961. Because many high school students have the ambition of being admitted to one of the highly prestigious Grands Écoles, the extremely challenging entrance examinations for them strongly influence high school curricula. In 1961, in response to the increasing interest in "modern mathematics", the committee responsible for those examinations put out a radically revised set of test-preparation manuals. Secondary education had no choice but to respond. Various textbook authors thus engaged immediately in this effort with collections that began basic set theory already at the first year of high school (C. Bréard, G. Papy), and a sort of race to modernity took place among various protagonists – among publishers, among private schools, among public schools, etc. The resulting rupture with elementary education became apparent, and stood as a challenge.

Could this reform go further down the education system? The principal obstacle was clearly the language and its use. The modes of definition of mathematical notions and establishment of knowledge seemed incompatible with the age of the students and the culture of elementary school teachers. The use of set theory resulted simultaneously on the one hand from the choice of very simple representations and on the other from the paradoxes and logical difficulties that this representation brought out. Teaching it incautiously would combine all the disadvantages.

The exercises of "modern mathematics" that could be conceived in accordance with the principles of the methodology of the period first put in an appearance as amusing additional activities without visible use for the acquisition of the traditional basics, to which they also added nothing. On the other hand, the work of Piaget showed children spontaneously acquiring mathematical structures with no need for knowledge of the terms describing them, or of explanations, or even of the essential organization of the mathematical discourse.

It seemed to me it was impossible to imagine that teaching primary school students could be done in the way chosen for secondary. Some people, later on, hoped to explain the fundamental terms to them and invent illustrative exercises, but what could be said to teachers absolutely could not be said to students. Others proposed metaphors and graphic representations and others finally suggested introducing 6-year-olds to symbolic logic. Attempts to make things explicit immediately put into play difficulties and paradoxes from which it was already clear to me that the teachers would not be able to extricate themselves. But the movement was too strong to avoid this obstacle. It needed to be channeled as quickly as possible.

The crucial questions, it seemed to me, were the following:

(a) Could fundamental mathematical concepts be taught to young students directly, as practices, without any formal verbal explanation being necessary?
(b) Would these concepts then be operational, that is, usable and useful and not just reconstitutable as isolated pieces of knowledge?

Other questions followed, such as:

(c) Would teachers be ready to practice this teaching spontaneously after a training reduced to reorganization of mathematical content? If not, would it be possible to give teachers training that would enable them to practice this form of teaching?

(d) Would the *connaissances* thus acquired be usable for the early acquisition of corresponding *savoirs*[6]? Would these conditions in the end ameliorate the ultimate acquisition of the mathematical knowledge, classical or modern, that was or might be taught? Could the practices thus introduced subsequently be correctly formulated? Spontaneously? Or only in certain conditions? Could the practices of notions and their formulation be learned correctly, without justifications and without proofs? At what moment could these diverse forms of justification be united, and coincide? Could children use and learn a formal system without recourse to verbalizing it and making it explicit in the vernacular? I had already tried out and succeeded with a number of non-verbal lessons; I had even prepared a complex non-verbal program dedicated to the calculation of fractions (as linear mappings) which I fortunately never tried to use. Where could this type of process lead? Many other questions and many other speculations came up.

The reader will understand that my conclusions could only be very pessimistic … but still…

Still it seemed to me that this research deserved to be carried out, not with an eye to immediate development or action-research, but because of its scientific importance. This type of study was thus indispensable to prepare for the future. People's enthusiasm could be put to good use to create a scientific organism, or some scientific organisms, dedicated to the study of these questions. A center for research on the teaching of mathematics could open the door to a specific institute by preparing the research instruments and personnel.

This was the sense of the conclusions I presented in 1968: the need was for the creation of an Institute for Research for the Teaching of Mathematics. What material means would this project necessitate? Given the issues and conceptions of the period, adding a few teaching posts and supplemental support personnel did not seem excessive. The alternative was the retraining of all secondary teachers – a massive expense if undertaken.

Experimentation: How and in What Form?

At the time, the experimentation that was in vogue was of a very different nature from what I envisioned. It tended to consist of a lesson or a curriculum, an organization or some material or other, put into action in real conditions for experimental purposes. If the design was found satisfactory, it was likely to be adopted directly or at least to be shown as an example, diffused and reproduced. The essential elements were the novelty of the design and the fact that it was ready to be put widely to use as it was or with minor improvements. At the time, this was the principal accepted means of attempting to improve teaching methods. To accomplish the reform of teaching, the only route envisaged consisted of asking confirmed teachers to learn

---

[6]For a discussion of the distinctions between *savoirs* and *connaissances*, see Chap. 5.

the new mathematical concepts themselves and then to conceive of ways to teach them and demonstrate their value experimentally. This conception was in fact a fools' market: the diffusion depended on outside factors and the success of a trained and convinced teacher offered no guarantee of the success of his emulators. Moreover, our ignorance of the conditions of reproduction of a design or even a teaching sequence, the label of "reference practice" attached to practices that are in fact heterogeneous, and above all the irrational importance assigned to the innovative character of the propositions being studied led me to suspect grave faults in this process, faults which were indeed subsequently to become evident.

The classic schema for carrying out this experimentation included some or all of the following steps: Prior *theoretical considerations* (1) making it possible to envisage an *original didactical realization* – book, material or curriculum (2), which then required *training of the teaching personnel* (3) who were to take on the responsibility of *carrying out this project, the teaching itself* (4). The *observation* (5) of this teaching then made it possible to obtain "results" (6), and report them to the *methodology* (7) of teaching which could subsequently furnish the facts for *theoretical reflection* (1).

These steps arose from six principal domains: *The discipline, pedagogical theories, teaching practices, teacher knowledge, methodology* and *methods of observation and evaluation.* Each of these evolved independently, each with its own dynamic of evolution. An essential element of their progress was borrowing results or techniques imported from other domains such as philosophy or psychology, the imports often being of dubious value, because their original discipline did not have the means to guarantee the conditions of adaptation to teaching phenomena, and because conclusions about the resulting developments were impossible to draw.

This kind of experimentation made it possible to observe the results and report them, but in conditions that included numerous choices that the experimenter had made in each of the domains, with each result conditioned by all of the choices. No general conclusion was possible. An experimentation of this nature chained together the choices in every domain and thus furnished a complete "product", to be judged globally and reproduced, without its being known whether the product was in fact reproducible and if so under what conditions. Effective experimentation requires a consistency that the collection of independent sources of decisions could not supply.

An alternative was the *spiral method,* consisting of a sequence of experiments conducted *by the same team* in such a way as to permit each repetition to use the consequences of the previous experiments to *improve on the elements of each domain that contributed to its composition.* It was a familiar format for the teachers, who were accustomed to presenting the same notions to their students repeatedly in the course of their schooling in order for the students to make progress on each one, profiting from their progress in others. But this time it was applied to the activities of the teachers and researchers and not to the students themselves. A repeated experiment conducted by the same team of researchers should create a much stronger relationship with the domains in use. Thus the spiral method could improve the product. But could it also improve the domains that had contributed to setting it up and thus contribute to the improvement of knowledge about teaching within these domains? For that it was necessary, but probably not sufficient, to insure at each step a collaboration of the experimenters with the specialists in each domain to compare the results from each other's points of view. This was the model that we retained for the functioning of the teams of the COREM.[7]

This could only function on condition that each time around the spiral worked well and provided actual progress in each domain visited. This empirical and pragmatic optimism was encouraging, but it had nothing to say about the nature of the efforts to be made on each sector of the spiral. If the successive experiments were not carried out by the same teams of teachers and observers everything depended on the quality of their capacity to communicate their conclusions. It seemed to us that the obligation to describe significant events without a new means of communicating them could only diminish and delay the progression up the spiral.

Later on we carried out experiments that demonstrated that teachers had difficulties in describing their activities precisely to their colleagues. This confirmed our views. Ordinary vocabulary lends itself to highly diverse interpretations. Repetition by the same actors would permit a considerably livelier rhythm of discoveries. What was needed was thus experiments repeated by the same teams at a rapid rhythm and continuing for a long time. Furthermore the premature diffusion of the observations outside of the circle of the research teams would be very likely to be fatal to the continuity of the initial conditions. The experimentation thus needed to be as discrete as it was ambitious.

I was able to put together a composite group (university faculty members, high school and elementary school teachers, administrators) who were to constitute the core of an organism that would make it possible to carry out original but administratively supported experimentation. It was necessary at the same time to organize, make functional, and describe the rules for the intended organization. In 1967 this led to the creation of a Center of Research *for* the Teaching of Mathematics (CREM), whose official mission was to be an organism for documentation for the teachers. This realization responded in part to Lichnerowicz's request (administration, organization, functioning, staff and personnel to get it started, orientation and methods,...[8]) It was accompanied by actual propositions for trials, preparatory research, research projects, methods, projections of methods, and elements of theory that would permit the proposed research organization to function. They were revealed in 1969–1970 in the first theoretical presentations.

Our Experiments

In contrast to the experimentation aiming directly at practice and the preparation of new developments, our aim was emphatically that of *studying* a phenomenon.

---

[7]The reader will find a description of the application of the spiral method in the organization of our research at http://faculty.washington.edu/warfield/guy-brousseau.com/files/possib711.pdf

[8]Readers will find the founding text of this organism at http://guy-brousseau.com/le-corem/le-crem-projet-de-lirem-de-bordeaux-et-du-futur-corem-1967/

This is a completely different challenge. Studying a phenomenon requires that it be identified, determined, and given a form, and then that it be questioned, seen to vary systematically or made to do so, with the help of observed or commanded variables. There was thus a need for a "theory": a sequence of definitions and declarations, methods for construction and aggregation of new statements, etc.

Next we had to produce some questions, and imagine designs that would make it possible to test these questions. In fact, the designs only existed thanks to the choice of a considerable number of conditions and hypotheses, organized into coherent possible models. But I also had reasons to doubt our capacity to examine the activities of teaching and learning without being prisoners of *a priori* justifications insufficiently concerned with reality. It was a matter on the one hand of the utilitarian approach (notably the requirements of promoters, publishers, politicians, etc. dependent on the short term affects on opinion) and on the other hand of the intellectual approach: each discipline had many legitimate propositions and conditions to validate, but combining them risked ultimately constituting only an insurmountable obstacle to all realization and all reflection on the object itself, which were our objectives.

To avoid all these obstacles I enlarged the spiral method to a scientific project, with the theoretical episodes connected through real activities. It was a matter of reconciling what was realizable with what was conceivable at that time and later.

The philosophical and epistemological bases of my approach seemed to me simultaneously natural and very novel. It was only far later that I had access to the original texts behind the popularized versions from which I was working.[9]

From Experiments to Theories: And a Science?

The pragmatic route was easy to conceive, if not to set up, but envisaging *scientific* experiments, in the sense of the period, as I believed Lichnerowicz had suggested, brought up interesting difficulties: no serious researcher in any domain would have dared to envisage the problem that way without being instantly discouraged, as much by the difficulties attached to each envisigeable angle of attack as by the opinions and objections that he would have to face from each discipline.

But why not apply to experimental research the principles of empirical experimentation? Why not envisage a spiral process for the development of a science that didn't yet exist? It would suffice to repeat in each sector of the principal spiral the cycles of work forming the sub-spirals. For example, one could begin by using light statistical methods to prepare for more sophisticated ones on better chosen observables, with more stable lessons.

---

[9]For example, I only discovered Wittgenstein's *Tractatus Logico-philosophicus* well after its translation into French (1961) while my positions – so close to his – were already strongly involved in my experiments. The same applies to Bachelard's notion of epistemological obstacle. I think that the necessity of conserving the logic of my approach led me to delay searching for the confirmations that I would have been able to find from more prestigious authors.

We determined a model to describe or envisage the activities of the students and the teacher in our experiments. This model permitted us to check whether the experiment was of interest for developing lessons and relevant for the study of their results, and to test our capacity to use them, analyze them, and possibly reject them. The principle was to start with the elements available to us and substitute new ones for them as they became available. If we had not yet succeeded in making certain experimental lessons satisfactory for use with students at the moment when they had to be given, they were simply replaced by classical lessons that were easier to prepare. Thus no method, pedagogical or other, was excluded in a general way. Only a priori and a posteriori analyses established the conditions and events associated with the lessons.

Otherwise stated, it was a matter of preparing teaching experiments on simple but specific questions, based on rudimentary but appropriate theoretical conceptions that furnished plausible criteria and methods for learning from the experiment. It was above all necessary that these experiments generate on the one hand new information and questions in order to pursue the process, and on the other hand improvements of each of the components of the spiral. The first question needed to call for the following ones.[10] The first realizations needed to permit the improvement of the functioning of the system, to teach the participants in the process (the teachers and the researchers), to sharpen the theoretical and experimental instruments – all in a way that would be convincing for the institutions that supported the system.

The Framework

From 1964 to 1970, the study of the questions listed above consisted chiefly of creating "frameworks" for principal elementary mathematical knowledge that could be envisaged for elementary school. These studies corresponded well with the missions of information and coordination for which the administrative organism that housed us (the CREM) was responsible.

A framework consisted first of an exposition that was purely classical mathematics, but complete, of the reference knowledge that was to be taught, in an order that prefigured the definitions, the theorems, the proofs, and their relative positions. Next came drafts of a series of fundamental lessons that would constitute the framework of the curriculum, completed by that of certain intermediate lessons.

Every lesson draft was characterized, more or less precisely

- By its *mathematical subjects*: an inventory of the notions involved in the lesson (definitions, properties, statements)
- By *general intentions*: what the students were to understand, to learn, to learn to do, to learn to say, … which would become more precise in the course of the preparation,

---

[10]Subsequently I often used the following test when a student proposed a subject for study: What question do you wish to pose? Do you have reasons for doubt about every possible response? What will you do with the response? (Let's assume that the experiment has been carried out and that the answer is known – what are the consequences?)

- By an inventory of the *didactical status* of each of these notions before the lesson, ranging from an *implicit notion already encountered but not identified* to a *notion that will serve as a reference in the course of the final learning* or a *familiar notion already acquired but revisited.*
- By an inventory of the hoped-for modifications to these statuses in the course of this lesson (a before and after pair for each status)
- By the type of lesson envisaged: classic (title, presentation, examples, questions, explanations, exercises, a problem), introductory or discovery lesson, review, game, etc.

This vocabulary is that of the period, but it was swiftly replaced by other terms: sequences of Situations articulated according to a process, questions spontaneously connected by a process: a logical thread (ascending or descending deduction) or a poïetic one (a spontaneous story as a function of the events.)

- By the place and role assigned to each of the notions relative to others that precede or follow it in the curriculum, and that this lesson cause to evolve.

We studied many frameworks at the same time for the same sector of mathematics. We gave ourselves the liberty of breaking with classic routes in order to examine them.

Nor did we hesitate, within the same curriculum, to break with the traditional discursive order to try to skirt or skip intermediate steps, or to invert the purely deductive order (which later led to "complexity jumps" (Brousseau, 1997, pp. 86–104, 174–176) and the notion of obstacles.)

These frameworks could have different rationales: classical or modern *mathematical order* (that is, the order of proofs and deductions) as in Chap. 2, or "poïetic order" (the order of a story, generated by questions) as in the introduction of statistics and probability (Brousseau, Brousseau, & Warfield, 2002). Statistical analysis of sequences of student results would then permit subsequent study of the characteristics of the lessons and their sequences.

In this way we began in that period to study a variety of rather original frameworks for the introduction of the designation of objects and collections and their enumeration, then for counting, numeration and the arithmetic operations. Logic, decimal numbers, measurement, rational numbers and geometry were to follow.

The first experiments that I undertook before 1969 were about the ergonomics of numerical calculation. My experiments allowed me to show the benefit of modifying the methods of calculating multiplication and division to improve the results and shorten the time required to learn them.

A framework makes it possible to make studies a priori and detect difficulties, inconsistencies, errors, and impossibilities, and to organize the experiments and studies that it might be most useful to undertake. These programmation processes were in no way original relative to the practices of the industries of the period. What was original was using these methods in the teaching domain and pushing them closer to the action of the teacher and students – also pointing out the doubtful points, the questions on which future realizations depended, the subjects of our studies, the provisional choices, etc. We were able to organize our progression in a spiral.

Observation

Can a lesson be considered by an observer as a ready-to-be-observed phenomenon? Class observation was practiced ordinarily only for training teachers or for the monitoring of their work by inspectors. Our prior conception was that class could be the seat of phenomena in the domain of different sciences: linguistic or psychological or social phenomena, or some other, or a combination, but not that it would itself be a phenomenon. That was a task! The first observations with a scientific experimental goal appeared at the beginning of the 1970s in the area of linguistic or social psychology studies (Flanders, 1976), but it was always a matter of studying a phenomenon defined by another science, studied in scholastic conditions. But what phenomenon was the teaching itself? The idea of carrying out anthropological observations of classes was inaugurated, I think, after the initial conception of the CREM (1967–1970), at the time when the COREM was getting started in 1972–1973.

Reflections on This Ambitious Project

Work done within the CREM did not result in scholarly articles, but it permitted us to go right into action from the moment of creation of the IREM in 1969 (see Part II below.) The bases of the Theory of Situations were presented in 1970 with an example of a framework (Brousseau, 1970). And we set to work to find a school to create a real center for research and observation of mathematics teaching: the COREM, with its school and its materials and its network of correspondents. This phase took us 3 years of efforts but from the very first year of functioning of the Jules Michelet School of Talence we were able to put into experimentation there curricula issuing from the best of our frameworks on natural numbers and on rational and decimal numbers, and in 1973 on probability and statistics.

I have been asked to explain the kind of incredible confidence with which a simple elementary school teacher could pursue for 10 years a project that was so ambitious and so much above his condition. Without going into details about what in this marvelous period made it possible to realize this small miracle, I would say that I had two advantages.

The first was my position itself. I never had to aspire to the official positions that would have seemed to others to be necessary in order to make decisions indispensible to this project. I always had the good fortune to find generous and attentive scientific and administrative leaders who were able to understand my projects, to support them and to make the necessary decisions that I asked of them. Some people might think that I was particularly adept at convincing or at attracting, but that was not the case. This success was due to the intelligence and devotion to the public cause of all those leaders, and to the fact that there were none of the obstacles of personal ambition or fear of perilous adventures. The project was ambitious, attractive and adventurous, but no one could doubt that it was altruistic, and no one, especially not myself, could achieve any career or financial benefit from this communal adventure.

The second is more singular. I had a sort of model. I had in my head *Les Souvenirs d'une Vielle Tige*[11] (Odier, 1955), a book of memories written by Antoine Odier, an engineer who had participated in the development of flight from 1908 on, of whom the builder Gabriel Voisin says that he was "one of those admirable utopians full of ideas, full of realizations, full of dreams, full of realities, capable of conceiving of a machine to explore time." This astonishing book described with great verve different aeronautical experiments, but also the spirit and adventures of the pioneers of aviation. And I compared the situation of what was to become *Didactique* with that of the beginning of aviation, drawing from it ideas, optimism and precious information. Basically it was a matter of simultaneously following the route of the Wright brothers that allowed them to resolve methodically, alone, before anybody else, all the practical problems of aviation, and avoiding reproducing the social isolation that in the end led them (like Clément Ader) to a sort of dead end. It was also a matter of avoiding falling back on the empiricism of technique – that is, letting the availability of a technique influence our choices when we needed to focus more on our goals. We needed to find someone who would do for Didactique what Eiffel had done for early airplane designers, pointing out design weaknesses and the consequences of some of the technical choices.

The Foundations (1970–1975)

The IREM [Instituts de Recherches pour l'Enseignement des Mathématiques][12]; the Bordeaux IREM

The first of the IREMs were created in 1969. Paris, Lyon and Strasbourg opened in January, Bordeaux in October. The "R" officially signifies "Research", but the principal activity, or at any rate the one that could be immediately launched and was the most in demand was that of professional development of teachers to prepare them for the new basis of mathematics. The government allocated to this task many "supplementary hours". This created some debate. Teaching the teachers to teach the new mathematics required for university admission consisted, for all of the IREMs, of teaching them, or at least those of them who could follow it, in the classic university format, and leaving it to them, as a friend of mine put it, the work of "setting it to music", that is adapting it to their own classes.

However this project, theoretically conceivable with a homogeneous class of well trained teachers, failed to take into account the actual capacity of some among

---

[11]The Association of "Vielles Tiges" was a group of pilots whose flights had been officially recognized before 1914, and who thus participated in the take-off of French aviation.
[12]Research Institutes for Mathematics Teaching.

them to carry out the proposed updates.[13] On top of that, the available textbooks, which took no account of the pedagogical and didactical conditions to be satisfied, could not be used. The "R" in "IREM" was interpreted by the unworthy term "Retraining".

In addition, others took it as indispensible that the IREM take on the work of the pedagogical adaptation of the intended mathematical program and hence that they reserve some of their means for the study and diffusion of appropriate forms of teaching at every level.

Among that set, there were still two trends. One consisted of thinking that groups of teachers and young mathematicians together would have sufficient knowledge for the construction of courses that were appropriate for students and correct from a mathematical perspective. The boldest among them dreamed of adding to these groups a psychologist or a linguist. For this group, the "R" in IREM did indeed signify "Research", but in fact this term referred not to any precise method, but rather to an intention. For example, groups of the IREM of Lyons (one of the first three IREMs created) essentially supported the production of manuals for the students and gave information to the teachers in an ambiance of creation of mathematical texts destined for the development of an updated mathematical culture. The IREM of Strasbourg undertook epistemological (G. Glaeser) and statistical (F. Pluvinage) research on teaching. Each IREM added some particular thing or other to the panoply of communal actions. In Paris, the IREM had the means to engage itself in all directions. An Institute of *Didactique* of the Sciences created at the University of Paris 7 interested itself in the teaching of logic and in its use in the analysis of student errors.

The other trend was the one I proposed to the IREM of Bordeaux, which supported it. Without declaring it too openly, the IREM of Bordeaux had the ambition of preventing the predictable excesses by an effort of rigor in its actions and by the development of authentic research.

The more one approached the teachers' actual practices in an ambiance of individual innovations, the greater was the danger of seeing an expanding tide generated by all sorts of utopias, educational, epistemological or something else, that neither administrators of education nor the IREMs would be able to stem. There was no way to prevent this florescence. What had to be done was to forestall some of the excesses by providing suggestions and examples and by developing debates. The debates would require that critiques have substance and consistency, thus work of a theoretical nature. They would also foster coherence by having exchanges that were open to all.

---

[13] The project ran afoul of the actual sociological composition of the set of teachers. At every scholastic level, primary and secondary, adapting to the consequences of the war and to the needs of the revival of the economy and to the budget reserved by society for education had led to the opening up of a large proportion of teaching positions to teachers who had good qualities, but had not received the expected theoretical training in mathematics.

The IREM of Bordeaux thus committed itself

- Not to innovations (which were known to be of temporary character) even accompanied by demonstrative experiments (which always succeeded but whose replication and generalization almost never did)
- Nor even to "scientific" research on "development" which had no direct relationship with the actions associated with it (because it always dealt with questions raised by excessively theoretical ideas) and which would collapse as soon as it was imported and set against the realities of school,
- But by undertaking research that was simultaneously fundamental and experimental, not in order to settle the question of this reform, but to accompany the development of a genuine science of the diffusion of knowledge, based on the specific requirements of the knowledge taught. This could only be a personal project. It was essential that the program not be exposed beyond the small group of people who were directly involved. Each action undertaken needed to be justified on the basis of the knowledge of the period and starting with its actual immediate and short term usefulness. But that is the project that is at the heart of all my actions. It was only several years later that the set of concepts that I introduced in 1970 at a conference of APMEP (Association des Professeurs de Mathématiques de l'Enseignement Public) was identified under the name of Theory of Situations.

Contrary to the habits of the period, "the" theory was not designed to be accepted as true and to be used to guide practical realizations to be diffused immediately in all classrooms. On the contrary, it was a model designed to be faulted by "closed" experiments and by debates and replaced by a less fragile model, until in turn new facts or arguments destroyed that one. The scientific advantages of this approach and its disadvantages with regard to the media leap to the eye, but its richness cannot be doubted. It made it possible to create a genuine scientific cooperation in a new field by focusing on the precise conditions observed or realized and leaving somewhat in the background the usual debates on education.

The enormously inconvenient aspect of this method was the following: The rapid advance of results and options isolated the first researchers from the new candidates for research. To anyone who had not followed the thread of the experiments, their results appeared to be ideas that were interesting, but fragile and difficult. Their coherence quickly imposed on new arrivals a work of initiation that was intense, painful and ultimately dissuasive. To be convinced, one need only consult the program of theoretical and practical courses required for a DEA in *Didactique* of Mathematics from Bordeaux, a degree created in 1975. On top of which there was all the work of experimentation at the COREM! We will give an idea of some of the concepts brought out and taught in Chap. 5.

Thus the team had prepared itself since 1965–1966. The Center for Research on Elementary Mathematics Teaching of the CRDP of Bordeaux assembled regularly and by contract more than 50 people who devoted to the project some part, small or large, of their free time or of their regular job (of teaching at elementary, secondary or post-secondary level, tutoring students or training teachers, among others.) Preparation began as soon as the IREM was created with the group assembled at the

CREM of the CRDP. It took 2 years to be able to create the required conditions. In 1972, the Jules Michelet School was able to open its doors to children in a blue collar neighborhood whose families had been displaced from the center of Bordeaux by gentrification. The agreement of collaboration of the National Education and the University of Bordeaux and with the minister to make the COREM official was never signed, but it was respected and financed by all parties for 25 years. Six experiments among the frameworks prepared over the years were immediately put into action: two at the pre-school and kindergarten level on the identification of objects and collections and the construction of a "code" to designate some of them, and on counting; two at the second and third grade on the calculation of arithmetic operations; and two at the fourth and fifth grade level, one on the introduction of decimal and rational numbers and the other on the introduction of statistics and probability. The system of observation, analysis and research that we will describe in more detail later was put in place in this period.

This project led me to create simultaneously an organism focused on the realization of experiments and research and a system of organisms and relations adequate and indispensible to guarantee its existence and survival. I had to:

- Supply the different groups – those doing training and those doing research – at the different IREMs with suggestions and texts ready to be used from their particular perspective and to their benefit;
- Attract complementary collaborations;
- Carefully reduce divergences, etc.;
- Do enough different forms of research to give credibility to the ones that we would suggest but not carry out;
- Do enough experiments to nourish reflections that would not turn into dogmatism,…

This part of my work, which I thought of as subordinate, was the most adventurous, the one that asked the most of me in efforts, in volatile inventions, in hazardous gambles, in disappointments and successes, and that brought me the most joy. I can't possibly give the details of it, but I still have the feeling of having been right in wanting only to do what logic and necessity required.

- Take the teaching of mathematics not only as a scientific field explored by different sciences according to their own methods (the position of my friend and accomplice the psychologist Gérard Vergnaud[14]) but above all as an autonomous body of concepts supported by mathematics, in some ways a science itself.

The experimental curricula (among them the one on rational and decimal numbers) were the essential instrument of this approach, and the COREM and the Jules Michelet School were my Galapagos Islands, if I may be excused the audacity of the reference.

---

[14]Thanks to whom our works were able to progress and to be funded by the CNRS (National Center of Scientific Research), and with whom I had the good fortune and the pleasure of collaborating for many years.

The COREM (1973)

In the previous chapter we presented, from the perspective of the teachers, the most important and original piece of the COREM entity: the Michelet School. But on the part of the researchers the originality of the arrangement and the enthusiasm of the participants was no less. The organization of the steps of each piece of research and of each piece of experimentation was precise and complex, but not rigid. The rule was a sort of tempered pragmatism. Each project unrolled in a precise order:

- Study of the different mathematical options,
- Translation of the "progression" into a framework,
- Study of the fundamental Situations that could generate the critical steps,
- Possibly, development and experimentation of certain of the choices made,
- Experiments "out of context" on precise important points when it was possible,
- Comparison of properties: the advantages and difficulties produced by the framework from all of the points of view (students, experimenters, teachers,...) and drafting of the curriculum,
- Carrying out and observation of a preliminary draft,
- Preparation of the didactical notes (precise description of the scenario for the lesson, texts of the propositions, expected reactions, etc.),
- Preparation of the teams for observation and recording of the class, of the behaviors of certain groups of students, and of the students,
- Collection and analysis of the documents produced by the students,
- Tests
- Evaluation and statistical analysis of the results of two classes.

The team responsible for each of these steps was composed in part of researchers who followed all the steps of the process, in part of specialists or people responsible for a step or a part of the process, in part of "moderators" external to the research whose job was to support the teachers in their ordinary work and in their debates with the researchers when there were any. Each function had its rules, and nothing was supposed to abuse the advantages conceded to the research. The sessions of observation and analysis gave rise to seminars, to lesson or project plans for sharing, and possibly to courses if there was an interest.

This way of functioning was respected for long enough to create habits that survived outside events: arrival of "new researchers in training" present for a short time, restriction of means, etc.

Further Developments over Time

The Diplôme d'Études Avancées de Didactique des Mathématiques (DEA) 1975

The "amateur" researchers who constructed the theoretical, experimental and methodological bases of research in *Didactique* conceived and led by mathematicians (with the collaboration of various other specialists) needed to train new researchers.[15] The means were given to them in 1975 by the authorization of three of the IREMs (Paris, Strasbourg and Bordeaux) to award a DEA (Diplôme d'Études Avancés – roughly the equivalent of a Masters degree) of *Didactique* of Mathematics to people who had received a DEA in Mathematics while following complementary studies within their Mathematics Department. At Bordeaux, the two degrees were soon combined.

The Doctorate of Didactique of Mathematics, Part of Mathematical Sciences

The sequel seems natural. The first doctoral theses in *Didactique* were presented in 1982 by young DEA recipients in mathematics who recognized their ability to carry out respectable mathematical work and also to teach mathematics to certain mathematics students at the university level. This type of thesis sanctioned theoretical and experimental work in *Didactique* of Mathematics and entitled the author to apply for jobs for researchers and teachers at the universities ... once the mathematicians were willing to give jobs to people with those credentials.

This measure crowned our efforts – those of my companions in the adventure and mine. Very few of us organized these experiments, this research, and this training with the intention of making a career of it themselves. The others were interested, like me, in the students, their teachers, and their harmonious acculturation to mathematics. I considered therefore that the task I had assigned myself 20 years before, and for which I had been given so much help and so many marks of confidence, had been accomplished. The moment (1979–1980) was a turning point for me.

At no point had I anticipated myself engaging in the career whose birth I had prepared. I thought of myself only as someone very knowledgeable about teaching practices, a "didactical engineer", able to set up the experiments the researchers needed, a good explorer of the territory that others had the legitimacy and power to organize. I didn't satisfy the requirements that I had set up. At the urging of my companions and my "students", whose affection and aid manifested themselves forcibly, I ended up agreeing to join them and continue to work with them, and to deal with the inconveniences of a position that many would regard as false. I had tried, twice over, to put together some of my work as a thesis, first with Pierre Gréco on the learning of natural numbers and operations (Psychology), then with H. Touanet (Statistics) and P.L. Hennequin (Mathematics) on the teaching of probability and statistics. But feeling that the work did not contain what I felt to be essential I had abandoned those projects. I took up the route again with the encouragement of B. Malgrange and in 1986 presented a somewhat composite but acceptable thesis.

---

[15] A graduate degree in *Didactique* of Mathematics.

Documentation

Most of the observations that illustrated and supported the ideas that I had before
1970 were made between 1970 and 1975. Few personal publications remain from
them. After 1975 the traces of my personal work can be found in the production of
my "students" and are reduced to suggestions for research, reflections, extensions,
repetitions and critiques. I am making as much of my written work as possible available at my web site: www.guy-brousseau.com, which is mirrored in English at
http://faculty.washington.edu/warfield/guy-brousseau.com/index.html.

Copies of student and teacher papers from the COREM can be found at a center
set up for them in Castellon, Spain, the CRDM-Brousseau of the IMAC of Castellon
(http://www.imac.uji.es/CRDM/index.php). Finally, although our choice of "home"
movie cameras to cut the costs of filming our lessons and our work sessions resulted
in the irremediable loss of more than 400 hours of records from the period between
1975 and 1988, there are nonetheless many video sessions available through ViSA
(Vidéos de Situations d'enseignement et d'Apprentissage http://visa.inrp.fr/visa).

Research Organizations in Didactique of Mathematics

Some structures have evolved over the years as *Didactique* has developed:

* The National Seminar of *Didactique* of Mathematics, held every 3 months, and
  a place for debate for all the teachers and researchers in this new domain,
* The *Association pour la Recherche en Didactiques des Mathématiques* (ARDM).
  This young scientific society brings together professional researchers in this new
  domain and their students. It devotes itself to the legitimacy and means of autonomous action on all subjects relative to its domain,
* Summer Schools in *Didactique* of Mathematics every second year, a place for
  deeper exchanges among the researchers and teachers interested by our efforts.
  Organized at the beginning with the help of the IREM and the minister of education, they evolved into a highly decentralized system that functioned and evolved
  autonomously under the responsibility of the ARDM.
* In France, the researchers are trained and chosen by Mathematics Departments,
  but their work is carried out in various composite laboratories[16] depending on the
  university.

---

[16]French universities have a structure that differs somewhat from that of American universities.
While the arrangements for carrying out the teaching function of the university are carried out by
departments organized by field (for instance, a department of mathematics), there is an almost
independent grouping by research focus. These groups, or laboratories, may be subsets of a particular department, but can also cross departmental boundaries.

The Current State of Didactique of Mathematics

The development of *Didactique* is currently blocked at the level of training of researchers. This training has become too short to transmit the results of 40 years of specific research. Various difficulties have dried up the supply of teachers, of diverse personnel and of organizations that held and supported the work. The work is tending to fragment and to align itself with methodological practices imported from other sciences and ill adapted to our subject.

Finally this science furthermore has no utility, because after years of an agony deliberately organized by a succession of governments, the specific training of young teachers has finally been officially eliminated from university teaching in France. Today, amidst general silence, school is the first field of battle and the first victim of the fight for power and for profit.

In all of these foundations, I was only one actor amongst many, too numerous to be cited here. But I feel the failures of this adventure as personal failures, even though they should be understood and explained by other factors.

Further Commentary on Professor Lichnerowicz's Challenge

What exactly did the subject Professor Lichnerowicz gave me mean? I have actually never ceased in the course of my career to learn new necessary conditions for the existence of an experiment in mathematical "pedagogy". An "experiment" requires a "conjecture", that is, an "alternative", submitted to a confrontation with a determined actuality through use of a precise experimental design. In teaching materials for mathematics, the combination of a precise and heavily structured text of knowledge and rationally established didactical principles left little place for alternatives. The uncertainty of the system was essentially restricted to the behaviors of the students and the teacher. And since an acceptable behavior or a correct answer gave no handle on the situation, all research seemed to be condemned to focus on predicting, inventorying, and explaining errors. This left the option of studying either the student or the teacher, and finally psychology for the former or misapprehensions about psychology for the latter.

The subject that Professor Lichnerowicz had proposed for me involved in fact – as I was swiftly to discover – conceiving of the teaching of mathematics as a new and real field of scientific research. The development of this project as such was to occupy more than 40 years of my life.

Institutional Difficulties

Until the 1970s, projects of experimentation on teaching in France ran into an obstacle of scale. Every citizen of age 6–14 was required to be educated on what served as the common base for relations of all citizens, and the government was to see to it that each of them had available the necessary means, that is, a teacher.[17] The government was responsible for guaranteeing the equity and effectiveness of the teacher's service. The former was obtained by making a teacher available to each student and imposing on that teacher detailed and uniform instructions, programs and schedules. The latter was attested to by the fact that a some of the students in each class achieved honorable success in the following class. If a large portion of the class had learned to read, then those who hadn't managed to do so simply hadn't taken advantage of opportunities that had indeed been procured for them. The only way forward was through regulated, executively declared texts. In other words, any modification had to be something that could be included and applied immediately and uniformly. The only concession to the art of teaching was that teachers were allowed to choose among a variety of theoretically unrestricted textbooks, but they remained responsible for the conformity of their work, so only books that were "consistent with the instructions" could be sold in quantity.

Difficulties in Experimentation

Experimental designs were practically all based on the supposed superiority of certain general educational principles like favoring the autonomous activities of students or individualization of teaching, or on more particular techniques. Global "successes" or "failures", even the spectacular ones, could never be attributed to precise isolated causes, so that the failures could never be "explained".

Possibilities for Experiments

Presenting a teaching sequence as an "experiment" in the pedagogy of mathematics required that one conceive of a hypothesis and a *reproducible* experimental design presenting at least two possible outcomes among which at least one contradicted the hypothesis and another would be compatible with it. But teaching is chock full of such occurrences, and the play of intentions and afterthoughts of the protagonists makes the methods of the "hard sciences" a priori illusory. Between the general conceptions and the precise observations are interposed a multitude of conditions and possibilities that are apparently impossible to make precise, to fix, and to reproduce.

---

[17]Translator's note: This is a place where English lacks a nuance: French has a separate word for an elementary school teacher: "*instituteur*", which carries with it the implication of instituting children into a society and not simply supplying them with academic knowledge. Currently, according to Brousseau, this aspect appears to be sliding into eclipse, but historically it is key.

Furthermore, the only possibility for slipping an experimental episode into an ordinary school was to reduce it to one or two sessions and place it judiciously in the midst of the course. Could one conceive of scientific experiments in these conditions, and if so which ones? In mathematics, the attempts had chiefly to do with materials favoring representations, manipulations and explanations. Piaget's designs offered interesting perspectives from this point of view, on condition of bypassing Piaget's objectives for the designs.

A New Conception of What It Means to Teach

Of what does a didactical action consist? The following definition was proposed by the psycholinguist M. Brossard to students of the DEA of *Didactique* in 1975: "A social project of causing a student to appropriate a *savoir* that is constituted or in the process of constitution." This clever definition is one that at the time – and undoubtedly also today – could be most easily agreed to by all interested parties. The choice, for example, of "appropriation" allowed equally for a Piagetian assimilation-accommodation and a classical teaching method, or even a behaviorist process. *Savoir* "in the process of constitution" presents a difficulty: how could it be legitimate to teach in a social project a *savoir* that would not yet be accepted as a reference? This definition was to be cited for a long period in my texts, but with a certain reluctance.

Because starting in 1970 I presented the object of my research as "a process of mathematization of the spontaneous activities of the students." Otherwise stated, it was a matter of provoking the spontaneous evolution of practices, uses of terms and modes of argumentation toward the corresponding mathematical activities. To be sure, this definition did not aim to describe directly my research at the period, which dealt with the teaching of basic knowledge to students too young to identify, designate and use the mathematical reference knowledge that corresponded to their actions and their learning.

Later, after having included situations of the teaching of mathematics at all levels as the object of modeling and research, and especially after finding in an article by Thurston (1994) a confirmation of my interpretation of mathematical activity, the definition became "a social project of acculturating students to the mathematical practices of a society." This new definition encompasses not only the teaching of reference mathematics, *savoirs*, but also the practice of their human and cultural environment: the search for questions, the questions, the treatment of *connaissances* and of conjectures, etc, a propos of the different aspects of the mathematical sciences.

Conception of Teaching

The classical conception of teaching is a formidable obstacle to its experimental study: it determines an ideal that is almost impossible to change: the Good. Failures are errors or insufficiencies: the bad. Alternatives are judged and received only as remedies relative to this absolute "good". Scientific study of this conception itself requires that one be able to construct detailed credible alternatives that can support hypotheses and their negation at the same time.

Construction of Alternatives

The classical construction of teaching (of all knowledge by and for all humans) rationalized by Comenius, developed in the eighteenth century, and institutionalized in the nineteenth century as a fundamental social function, appeared to be an undisputable, system, the ideal, the good, to which practical realizations were simply more or less defective approaches, suffering from ills inherent to this imperfect world. Any alternative ran immediately into a wall of objections. Failures could only be due to insufficiencies of the participants, first the student, then the teacher, knowledge very superficially (the foundational social reference of teaching depended only on the necessities of the discipline and certain rhetorical or didactical arrangements.) Timid attempts focused on the identification of difficulties and failures resulting from the standard practices, and offered only local rearrangements or radical utopias. Finding general alternatives to this system seemed impossible because they were imagined to be things that could be rapidly put into action and immediately diffused.

Now, the works of Piaget furnished this alternative: the development of knowledge of a child is a solid counterexample to classical didactics.

The Contributions of Piaget

A negligent reading of Piaget's theories induced some educators to await from a hypothetical spontaneous development of their students the fundamental learning that they hoped to obtain. But how could one believe that it would suffice to leave the young human brain alone in its current environment in order for it to acquire a culture and knowledge that appeared in exceptional circumstances in the course of centuries of patient work and chance events, 1,000-year hesitations, accidents, unconsidered choices and all kinds of alterations? The human brain does not have the power to imagine and recreate, alone, what humanity has created in the course of a long history – that is, its culture. What is transmitted at best are certain faculties for adaptation and resistance to adaptation in the face of unexpected conditions.

On the other hand, the experimental designs imagined by Piaget, directly inspired by his exchanges with the mathematician Gonseth, showed how experiments could directly reveal the presence or absence in the child of certain precise pieces of knowledge of a mathematical nature, in the absence of any teaching. But Piaget studied spontaneous development, and for that reason kept as far as possible from reference knowledge and school teachings. So it seemed to me that it would be possible to offer *situations* that could fairly directly provoke students to learn or correct knowledge by designs of this type appropriately modified. And I was even convinced that the study of these situations, be they carried out either spontaneously or systematically by the teachers or conceived as experiments, was a completely indispensible and nonetheless totally new object in the field of research. I could see very well what limited Piaget in the production of designs aimed at revealing fundamental structures in children's knowledge: it was consciousness that the designs – including the observer or teacher – had to be the primary objects to be analyzed systematically from the mathematical point of view.

The Notion of Situation

The notion of "Situation" generalizes and make more precise the classic one of "problem". It takes into consideration the fact that the student is interacting with an "objective" milieu that gives her part of her information (possibly in non-verbal form) and reacts to her decisions in a way that is independent of the interventions of the teacher, that is, following a (non-didactical) logic of its own. A Situation can thus model cases where certain conditions and certain responses are intentionally left implicit.

The notion of Situation also covers that of "design" in Piagetian language. The latter describes only what the experimenter has laid out. It is also necessary to be able to analyze the conditions of fortuitous happenings, of spontaneous learning situations. The use of Situations enabled me in particular to envisage interventions with students too young to use the language corresponding to their mathematical actions, and earlier than one thinks them capable of being instructed by a discourse. Or again to suggest to teachers exercises whose statements would have had to make use of knowledge the teachers didn't have or couldn't use (cf. Brousseau (1965)).

Replacing "problem" or "lesson plan" by "situation" thus led to replacing the classic conception of teaching with another. The classic plan consisted of the teacher maintaining a *teaching-discourse-for-the-teaching-of-a-(mathematical)-savoir*. Otherwise stated, he had to set up a (didactical) discourse on a (mathematical) discourse on the mathematical world and in short on the world. The new concept of "situation" made it possible to envisage *organizing for the student a certain direct mathematical experience of the world.*

This point of view was moreover made necessary by the fact that the verbal form of mathematical knowledge was totally impracticable in a class of young students, which prevented any specific commentary on the part of the teacher. The fact that the mathematical discourse at the time was new in the culture and dealt with the foundations of mathematical thought themselves offered an historic occasion for reinventing the bases of teaching mathematics that would not soon recur.

We had then to imagine situations that would cause a mathematical thought to be *born* even though it could not yet be formulated either by the student or the teacher. We had to imagine situations capable of making this thought *evolve* if it was incorrect by having it reveal itself to the student as inadequate in the situation for which she was responsible only for the procedure and the ending. The object thus identified and determined within a situation could then be *named* without having to be immediately defined and bounded verbally by a context that was still too narrow. Testing out this knowledge at the same time in its source, in its relationships with others and in its uses then became possible. This approach made it possible to conceive of a reorganization of relationships among the different elements of the cognitive function: the learning of a new behavior and its formulation, and its *constitution as knowledge* inserted into the midst of other pieces of knowledge. And this reorganization could be studied systematically, realized and modified. Beyond that, it made it possible to put each of these elements into a relationship with specific types of situations: of actions, of communications, of justifications, and of reference, and thus to conceive of modalities of learning differentiated according to these types of interactions.

These conditions explain the sentiment that there was no other way to go except to study these situations, spontaneous or organized, but always generators of mathematics. These situations opened up a route towards early knowledge, and carrying out a kind of very early teaching that I judged to be indispensible and possible. At the beginning I called these situations "didactical situations" as a contraction, because they were in fact "mathematical situations designed for didactical use", derived from mathematical situations which themselves had no didactical intention. It was only a little later that the ambition to study these models and their relationships in a systematic way led us to consider them as drafts of "a theory of situations".

Now in order for an alternative to be instructive, it was essential to examine its consistency in detail: examine the particular *savoir* that is the object of this exact exercise, in what precise circumstances; envisage its role and its alternatives; consider the functions of this *savoir* for every student; envisage its role in outside acquisitions, …. It was necessary at the same time to give ourselves the theoretical instruments for this precise examination. In this system every attempt at a general explanation had to be embodied in an intelligible concrete alternative of the situation sufficiently precise to make it possible to set up another similar and realizable one.

The *systemic approach* thus made it possible to isolate a reasonable number of coherent conditions and oppose them with an alternative system of conditions.

First Questions

The questions that were at the origin of my research were at first constructed on the same model and concerned all mathematical knowledge:

Is it possible to teach, or teach better, the subjects of primary school [logic, number, numeration and arithmetic, algebra, statistics, geometry, measure], *using,* and *staying as close as possible to* current mathematics? The teaching I had been practicing felt to me like a museum of the errors of humanity in the construction of its mathematics.[18]

How could students benefit from the new organization and recent progress in mathematics and various other domains (like genetic epistemology and linguistics)? What objectives could be assigned to that progress? What difficulties might present themselves? What limiting conditions must the educational system realize to allow the success of different types of reform of the practices of teaching mathematics?

The two questions that seemed essential to me were: *What situations should one offer the children to obtain a reasonable appropriation of mathematical concepts? At what age is it advantageous to begin this teaching?*

I suspected that my conception of the teaching of mathematics to young children was coming up against a completely erroneous ancient conception. Rather than following the example of the first learning of language, parents wanted schools to follow an academic and falsely rational path. For instance, that of the teaching of foreign languages in my childhood, where the languages were taught like Latin, by teachers who didn't speak them (or refrained from teaching them until they had taught their grammar and lists of words or irregular verbs.)

It soon appeared to me that these questions, fairly common at the period, rested on illusions and metaphors. Still, these two questions were direct consequences of the works of Piaget and his students. They spread throughout the teaching world, but they were interpreted in a variety of ways.

A Child and a Concept

Let us return to our two principal questions: what situations? And at what age?

Knowing at what age one can begin the teaching of a mathematical reference concept is not a subject for simple research. First one must have a concrete definition, or at least a model of "knowing a concept". One must be able to model the manifestation of a concept in the behavior of an individual or a collection of individuals (are there latent concepts?), and undoubtedly also have a model or an idea of what a concept is.

Models of a Genesis

Next one must have a model of the genesis of the concept, that is a double chain:

A *chain of pieces of knowledge* linked by *logical relationships* that establish the mathematical necessity of the concomitant (in a definition) or successive (in a proof)

---

[18] I like museums and monuments, but not living and raising children there!

presence of certain pieces of knowledge (statements) constituting the genesis of a terminal (new) piece of knowledge.

A *chain of situations*, of circumstances in which the pieces of knowledge determined by the chain of pieces of knowledge manifest themselves successively. In the traditional minimal schema, the sole objective of the situations is to establish the presence and permit the exercise of the successive pieces of knowledge. But I wanted in addition to have the situations make it possible systematically to illustrate and test the adequacy of the pieces of knowledge constituting the mathematical genesis. Otherwise stated, to have the situations introduce a different type of necessity from logical necessity for the pieces of knowledge: the type of necessity that guides mathematicians before they have established and made completely explicit the mathematical relationships that they are trying to establish and prove.

The double chain of pieces of knowledge and situations is indispensible for the reproduction and learning not of a text but of a transposed but active mathematical development.

The knowledge stages play a role in this of instantiation and functionalization of pieces of knowledge, that of reference, and that of accompaniment, whether one wants it or not, and of their succession. The curriculum is not an exposition, but a genesis.

The progression is made by logical necessity and/or by necessity produced by actual results.

This notion of double chain is at the base of my experiments on long processes at the period.

One could imagine that it is then possible, at least theoretically, to identify in this double chain a piece of knowledge and a situation that constitute the threshold that the child has reached in her development. And in the wake of this thought, to imagine that if all the children follow the same genetic chain, one could determine

- In the case of a spontaneous genesis, the average age of students who achieve this step
- In the case of a didactical curriculum, the cost of success, distinguishing two cases: that in which the speed of the progression is imposed and that in which it is "free", One could then observe the time necessary to obtain some fixed success rate for each step.

Clearly each of these hypotheses is a fiction. But, with refinements that conceal the fundamental error, they underlie popular reasoning as well as most of the assessment enterprises. Could they lead to a scientific development, even if only one of study?

The Standard Presentation of Mathematical Concepts

The classic example of modeling of a piece of mathematical knowledge is that of the standard textbooks of mathematics, of which the reader will pardon me for giving a more precise description:

Knowledge is presented there in the general order "definition, theorem, proof". It is accepted that the only knowledge that appears there is recognized and held to be *true* in the light of today's mathematics. The manner of establishing them it is codified: statements that appear there are constructed with the help of permitted constructions and of statements previously established. It is a fractal type order, to be applied to isolated statements and to groups of statements. With the result that every written statement is true – except for those made in the course of a proof by contradiction – and thus may be used as a *reference*. In reality, the text of mathematics presents a whole graduated variety of types of reference: definitions, fundamental theorems, lemmas, corollaries, examples, exercises, problems,... Similarly, the mathematical text is generally accompanied by two other sorts of text that, with it, constitute the "mathematical discourse":

- The meta-text of terms like "definition, theorem,...", line or statement numbers, the name of a mathematician recognized as the author of a theorem, etc.
- References to other scientific texts, commentaries, notes: connections with other parts of the text, counterexamples, historical remarks, etc. These carry an essential role that is not recognized as such, because it is at this level that the visible didactical efforts take refuge. The most important are nonetheless in the organization of the set making up the chain of pieces of knowledge.

Connaissances and Savoirs

These texts present what is considered to be the sum of the references to which mathematicians are held to refer in order to establish both the truth and the originality of their results. They thus institute "the *Savoirs*" – the reference knowledge of the community of mathematicians.

Connaissances are the traces of some relationship between a subject or a population and some object (material object, living being, entity, idea, etc.). *Savoirs* are the *connaissances* accepted as references by a subject or by a population. Different individuals or populations share certain *connaissances* and accept certain common references but diverge on others, sometimes without knowing it. Exchanges of *connaissances* are much easier than sharing references. In mathematics, the references – the *savoirs* – are the statements held to be true by the community of mathematicians, and the means of expressing these statements. They are reputed to be verifiable by anyone who has the necessary competences. In fact, there exist a whole gradation of "savoirs" determined by or attributed to a variety of populations, but not necessarily recognized by others. The *savoirs* are thus a very limited part of the *connaissances*.

Traces of any relationship of a subject with mathematics could be *connaissances*: first the specific objects of the mathematical language specific to the community of mathematicians: isolated signs, incomplete expressions, terms, well formed expressions, whether closed or open, true or false, theorems, whether references or not. But also metamathematical languages, the specific elements of mathematical and/or didactical discourse, the jargon of mathematicians and of different specialized professionals.

Next this enumeration must be extended to objects that make reference to the knowledge of mathematicians but as they present themselves in other communities of society.

We are interested in mathematical knowledge in the forms defined above and which teachers may find themselves confronting, but above all we are interested in the functioning of this knowledge in the process of acculturation of their students. We must therefore interest ourselves in what is the object of the teacher's work: developing the *savoirs* of their students and the functioning of the *connaissances* that nourish them.

In any process of creation or learning of mathematics it is indispensible to favor the play of *connaissances* that contribute upstream to the formation of *savoirs* before disappearing or being transformed. The process of learning is no other than this transformation, this specific debate, this dialectic between the known and the unknown.

The belief that only true statements participate in the mental process of production of knowledge seems to me contrary to observations. Reducing the production of *savoirs* to an arrangement of established *savoirs* is traditionally the avowed goal of classic teaching, but the goal of my theoretical research was to know whether this arrangement was optimal. I was dubious.

My way of expressing the process of acculturation of a student to mathematics followed a long route and went through many variations, hesitations, and modifications of language and point of view, but the object remained the same: the process of mathematization of the knowledge of the students in the course of mathematical and didactical situations.

The Place for Connaissances: The Types of Mathematical Situations and Theories

For 10 years I concentrated on ways of favoring the development of conditions for the functioning of three manifestations of *connaissances*: *implicit models of action* – *connaissances es that operate in a situation* independently of the possibility of formulating or proving them – *formulations* and *arguments*. Each manifested itself in a different model of situation, following different modes of acquisition. During this time I deliberately set aside the study of a type of situation that I later called a Situation of Institutionalization, by which the teacher introduced or established known *savoirs*

- Either because he revealed, posed or transmitted them himself, for example by assigning an activity, or in a classic "lesson" by defining and naming the elements which thus became provisional or definitive references,
- Or because he identified them, among those proposed by the students in the course of an activity, as being the expected result or *connaissance*, the canonical form, the reference for the pursuit of the quest and the production of new means and new *connaissances*.

It seemed to me essential to distinguish the principal object of our research: the functioning of situations that simulated and stimulated an activity of mathematics itself, from those by which the mathematician, like the teacher, must communicate her results and let the community share them.

I did not try to analyze the latter, but I believed I knew them well enough. Which is not to say that I rejected the use of them. Very much to the contrary, the standard organization of lessons was recommended to the COREM for lessons that were not the object of experimentation and that we used as an emergency procedure in case an attempt was considered to have failed. Every didactical situation that has ever existed and been used has a domain of conditions in which it is optimal, if not it would not have existed. None is optimal in every situation. Thus quarrels based on general principles are sterile and dangerous.

Much later, I separated the *Theory of Didactical Situations*, resulting from my 1980 reflections on the paradoxes of the didactical contract, from the *Theory of Mathematical Situations used Didactically* that I am in the process of describing. Comparing these situations with the functioning of the society of mathematicians I ventured to find them very close to each other, and I boldly rebaptized the latter the "Theory of Mathematical Situations". My hope is that this somewhat provocative title will generate both discussion and study.

Chapter 5
Expansions and Clarifications

In this chapter, we have chosen a few of the concepts and issues of *Didactique* that we feel are worth a somewhat more detailed discussion than we were able to able to give them in the previous chapters. This is neither an encyclopedic list of concepts nor a maximally deep discussion of any one of them, but rather an attempt to clarify a few key ideas for the reader.

Connaissances and *Savoirs*

An important characteristic of this curriculum is its handling of the two essential forms of knowledge: it distinguishes between the acquisition of *connaissances* and the acquisition of *savoirs*, and respects the roles that these two forms of knowledge play with respect to each other.

Events in class have the effect of provoking students to react, make declarations, reflect, and learn, all of these manifesting their intellectual activity. This activity reveals their *connaissances*: what they do, their intentions, their perceptions, their decisions, their beliefs, their language, their reasoning.

Only one part of this set of *connaissances* is recognized as expressible, and expressed, whether by the student, by other students, by the teacher, or by society. These *connaissances* are recognized with the help of a repertory of *reference connaissances*: custom, language, rules of orthography, established definitions and theorems, logic, communal beliefs, culture, etc. These are the *savoirs*. *Savoirs* are the indispensable means of recognizing and expressing *connaissances*, just as metalanguages are the means of talking about languages. One student's repertory of *savoirs* may not coincide with that of another student, with that of the class, etc., but in communications the repertory needs to be common. All *savoirs* are accompanied by an environment of *connaissances* that make it possible to use them. *Connaissances* that are not connected to any *savoirs* swiftly disappear.

Learning, whether that of a student or that of a class, manifests itself in the appearance of new *connaissances* and of new *savoirs*. Their systems of functioning are very different.

Connaissances may be correct or false, approximate or dubious, conscious or unconscious. They control the decisions and consents of the protagonists: teacher and students. Only those identifiable with the aid of *savoirs* can be exchanged. But all can be modified and corrected by all sorts of causes and means, some of which operate on the *connaissances* (modifying their relationships), others directly on the *savoirs* (definition, demonstration, etc.)

For the same user, the status of "connaissance" and of "savoir" of the same statement can vary according to circumstances. For example, a theorem can be a *savoir* in the context of a course, but in another context be just a *connaissance* until the pertinence of using it in the new circumstances has been established.

Learning thus manifests itself also by transformations of status: *connaissances* become *savoirs*; the negation of a false *connaissance* can become a *savoir* or the *savoirs* can be enriched and invested in new conditions and support new *connaissances*.

Didactical Situations

Before we delve more deeply into this essential concept, we need to make a notational clarification: for a number of years after we began our joint work, in both that work and Warfield's independent work any technical reference to a Didactical Situation was capitalized. At the time, it seemed important to make it clear that we were not referring to a funny thing that happened on the way to the blackboard, but rather to something serious and deeply important. Over the years, as *Didactique* has become increasingly familiar to Anglophone educators, this cumbersome device has seemed increasingly unnecessary, and we have been phasing it out. Since this book contains a mixture of older writings revisited and new writings, it also contains a mixture of the two notations. What follows, however, is newly written, so it will be without the capitals.

Underlying the idea of a didactical situation is that of a mathematical situation. This is an essential element, because the whole theory is based on replacing the prevailing classroom view of mathematics as a collection of facts and procedures with that of mathematics as an activity involving both *connaissances* and *savoirs*. Early on, before constructing the whole theory, Brousseau developed the hypothesis that causing the evolution of the imperfect or even erroneous knowledge – the "implicit models" – developed by the students could frequently be more efficient for a larger number of students than the direct teaching of formal reference knowledge. This evolution could only occur in response to a mathematical situation. In this context, he made the following definition:

A *mathematical situation* is a set of specific conditions in which a determined set of mathematical *connaissances* (stated or belonging to the *milieu*) permit a subject to realize a declared project by the exercise of appropriate mathematical *connaissances*, known or original. Note that this description applies not only to students but also to mathematicians.

The earliest definition of *didactical situation* was simply "a mathematical situation used didactically (serving to teach.)" As the theory progressed, it became clear that more specifics were needed. This need brought about the definition of an *a-didactical situation*. A-didactical situations occur in the classroom, and have the goal of reproducing the conditions of a real mathematical activity dealing with a determined concept: i.e., a mathematical situation. In the course of an a-didactical situation the students are supposed to produce a correct and adequate action or mathematical text without receiving any supplementary information or influence.

With this definition in hand, a *didactical situation* can be defined as the actions taken by a teacher to set up and maintain an a-didactical situation designed to allow students to develop some goal concept(s). In particular, the teacher sets up the *milieu*, which includes the physical surroundings, the instructions, carefully chosen information, etc. The *milieu* may or may not include a material element (for example Cabri geometry), and other cooperating or concurrent students, etc., but it does at the least include the *savoirs* of the subject and certain of her current *connaissances*. It is essential that the *milieu* be a design that obeys only "objective" necessities, and that the student be convinced of that fact. Once that design is in place, the teacher's mandate is limited to making sure the students focus on the *milieu* and not on the teacher.

Within the category of didactical situation, there are three notable subcategories, chosen because they correspond to models of completed mathematics or because they have an important place in the genesis of a concept. *Situations of action* reveal and provoke the evolution of models of action without the student's needing to formulate them. The student can, immediately or later, learn to identify them, to formulate them in *situations of formulation* (expression or communication) and to justify them in *situations of proof* (validation or argumentation.) There is a tight correspondence between (a) the composition and organization of the *milieu* (game, communication, debate), (b) the nature of the interactions of the subject with the *milieu* (action, formulation, proof), (c) the type of knowledge these relations call forth (implicit models of action, languages, mathematical *savoirs*.)

A fourth type of situation is that of *institutionalization*, which we will discuss further in the next section.

Institutionalization

At the level of the class, the act by which a *connaissance* becomes a *savoir* is *institutionalization*. Institutionalization modifies the rules for using the knowledge: for example by validating the solution to a problem. The student can then consider this knowledge as his *savoir* and use it in the solution of another problem. But in the didactical relationship, it may have to remain a *connaissance* if the student is not authorized to refer to it in the course of demonstrating the solution to another problem. The teacher attempts to increase the students' capacity to solve problems, but defines the repertory of reference theorems that they are allowed to use in their demonstrations. They are free to use their *connaissances*, but must not confuse them with their *savoirs*.

Institutionalization can be projected or realized as a didactical act or even declared, without in fact being effective. The final result of institutionalization is the appropriation of the *savoir* and its pertinent *connaissances* as obvious, as direct and commonplace expressions of thought.

The *mathematical connaissances* developed by the students in certain circumstances favor, in others, the acquisition of the corresponding *scholarly savoirs*, and vice versa. The conception and carrying out of a curriculum demands a rigorous respect for these interactions. The solidity and rapidity of learning depends on a day-to-day management that distinguishes meticulously what is said, seen, attested to, shown or hidden, agreed to and known. Teachers must be able to treat in the long term what it is at each instant that must be said, can be said, should not be said but can be given to understand, and shouldn't be said at all.

The scheme of things that situations of action have caused to be discovered, situations of formulation then cause to be expressed, situations of validation cause to be demonstrated, and other situations cause to be taken as a reference, to be institutionalized, to be studied and if necessary to be practiced. Finally yet others cause it to be considered as practicable by everyone and no longer needing to be said.

Didactical Contract

One of the concepts that was articulated early in the research efforts of *Didactique* and penetrated the language barrier just enough to cause a great demand for translation is that of the didactical contract. It has now been discussed in various English language articles, most notably the Case of Gaël in the Journal of Mathematical Behavior (Brousseau & Warfield, 1999). Herewith a somewhat compact description and discussion:

The *didactical contract* is the set of teacher behaviors expected by the student and the set of student behaviors expected by the teacher during a didactical situation – and in particular those specific to the knowledge to be taught.

Many questions arise when one focuses on this contract. Can these two sets of expectations be reconciled – be clearly formulated and negotiated? Are there divergences between the expectations that are in fact irreducible? Do expectations exist that cannot be expressed? What are the specific roles of what cannot be expressed, of what is said, of what is not said, and of what cannot be said to the student in the teaching relationship?

The situation is further complicated by the presence of a third partner, the client – the parent or society – who anticipates particular behaviors and particular results from the student and from the teacher.

These questions initially arose in research on the possibility of enabling a mathematical situation to take the responsibility for managing elements that the teacher could not say or the student could not yet understand. Two specific contexts brought them to the fore:

The case of Gaël: Gaël was an 8-year-old who kept answering mathematics questions in the manner of a very little boy. The researchers determined that the source of his behavior was not developmental delay, but rather a habit of avoiding risk by refusing any responsibility for what he said. As soon as lessons with "games" were included where the boy could take a risk and see the effects of his decisions and where he could get involved in betting – without a big risk – on the validity of his answers, the experimenters observed a rapid and radical change in his attitude and the disappearance of his difficulties. A new "didactical contract" with him had been created.

The Age of the Captain: The researchers from the IREM of Grenoble gave the following assignment to 8-year-old students: "There are 26 sheep and 10 goats on the boat. How old is the captain?" 76 out of 97 students answered "37 years old." This research caused general uproar. Either the teachers were accused of making their students stupid or the researchers of setting a stupid trap for the students. In a letter to the experimenters G. Brousseau suggested that it was in fact a typical effect of the didactical contract and that it was neither the teachers' fault nor the students'. The researchers verified this with a follow-up question to the students:

> What do you think about this problem?
> It is stupid!
> Why did you answer?
> Because the teacher wanted us to answer!
> And if the captain were a 50-year-old?
> The teacher didn't give us the right numbers.

An experiment with qualified teachers produced similar results: for various reasons such as the hope of provoking an explanation, the subjects produced the answer the least incompatible with their knowledge, even when they saw very clearly that it was wrong.

The use of tacit didactical contracts is as old as mathematics itself. In particular it allows construction of mathematics before looking for its foundation. This is why its importance and our ignorance on the subject became apparent at the time of the attempt to reform "modern mathematics". But, in response, the launching of explicit, naïve and vigorously imposed "contracts" (High Stakes Testing) reduced the sensitivity to the implicit contract, which only increased the difficulties.

The didactical contract manifests itself mainly when it is broken. Its experimental and theoretical study became crucial because the constituents' decisions no longer seem to be based on sufficiently appropriate scientific and/or cultural knowledge.

Connaissances and Epistemological Obstacles

The experiment that we describe in this book helped to bring epistemological obstacles to light and to study their affect of on learning. An epistemological obstacle in mathematics is a *connaissance* (not the lack of a *connaissance*) or even a *savoir* that has the following properties: it is valid and appropriate in a certain domain, but becomes inappropriate outside of this domain, generally without the inappropriacy being noticed by the person who wants to use it; it causes errors of unsuspected origin; it appears to resist all attempts to adapt or improve it locally; furthermore, even when a new and valid *connaissance* has been substituted for it, the epistemo-logical obstacle recurs unpredictably and creates incomprehension or errors, which makes it essential that the obstacle be identified and that its consequences be acknowledged. We have proved (Brousseau, 1983) that epistemological obstacles *inevitably* appear in the historical and psychological genesis of knowledge. There are numerous examples in the form of famous errors such as Cavalieri's generaliza-tion of indivisibles, or Lagrange's error on the simple convergence of continuous functions. The solution for the difficulties that they produce depends on a number of factors of the culture.

Epistemological obstacles make an appearance also in teaching, where they unavoidably get in the way of some necessary passages. The most astonishing but also the most obvious is that of the natural numbers. They are inevitably the founda-tion of mathematical learning and nevertheless they form an obstacle to some other *connaissances*. In particular, the fact that every number has a next number, or that the product of two numbers is greater than or equal to either factor needs to be aban-doned when numbers are extended to include rationals. Another obstacle that is pertinent to our experimentation involves division: students understand division in the natural numbers as expressing sharing in equal parts. Implicitly, the divisor and the quotient are necessarily smaller than the dividend. "Division" of two decimal numbers, even though its interpretation is different, is initially conceived on the same model. Students interpret $37.5 \div 6.3$ by analogy with $37 \div 6$. But since division by 0 has no meaning and division by 1 not much more, students have difficulty conceiving of $19 \div 0.8$, or even of $19 \div 1.8$. Much of the difficulty of rational num-bers comes from the fact that the divisor can be greater than the dividend. The dif-ficulty reaches its maximum with the interpretation of operations like $0.30 \div 0.80$. The models introduced with the natural numbers interfere in numerous errors even with advanced students, in particular in the study of Analysis.

But necessity or didactical fantasy can give rise to new, purely didactical obsta-cles. We consciously introduced fractions by commensuration (the search for a common unit permitting a comparison of two quantities) and we demonstrated that it did indeed constitute a (didactical) obstacle to the more classical comprehension of fractions as measures with a sub-unit. We were able to observe how the dominant conception could be imposed and how the students were able to use one or the other more easily depending on the case (but not both at the same time.)

The curriculum that we present is entirely conceived to treat or deflect the vari-ous obstacles created by "natural" conceptions to the comprehension and use of rational numbers.

Metadidactical Slippage

In case of difficulties or failure, the teacher is expected to intervene. Some of these interventions consist of continuing to make use of the situation under discussion without much change in the structure (furnish some information, ignore or correct an error, accept a weak response, etc.). Others break into the process: they cause the action to be abandoned in favor of another, different or similar (give a partial exercise subdividing the situation or the knowledge in question or a similar exercise in a different environment, revert to the explanations prior to the failure, etc.) Between these two modalities is another that consists of commenting on, explaining, discussing and studying the situation that is causing difficulties. This arrangement is supposed to insert itself like a parenthesis in the situation under discussion, suspending it long enough to obtain some useful information before the moment of returning to the original situation. But in fact it really is a new situation: its rules and means are different. The fact that the first situation and the knowledge that might resolve it become an object of study necessitates a repertoire for *identifying* and *writing* some of its components and properties (some, but not all – for example, the non-expressible knowledge of the children cannot be included.) This repertoire, itself made up of *connaissances* and *savoirs,* plays a role on this occasion comparable to that of a meta-language with respect to the knowledge in the initial situation. The situation itself is a meta-situation: its *milieu* is the initial situation and its rules are a priori specific. There exist a large number of types of meta-situations, for example:

- Those that can be classified by their objective: addressing the cause of the difficulties as being the student, the conditions or the *savoir* in question
- Those that can be classified by the means mobilized: the forms of argumentation (explanation, clarification, representation), the types of rhetorical device (metaphors, metonymy, etc.) of words, of thought or of discourse.

It can happen that this meta-situation fails and the initial situation cannot be resumed. When this happens, a new "meta-meta-situation" may appear to be a reasonably economical choice to save the whole activity, while abandoning the activity seems a pure loss.

In spontaneous intellectual activity, the passage from action *in* the situation to study *of* the situation is frequent, rapid and difficult to distinguish. In teaching, the distinction arises between the phase of problem-solving by a student and that of the study of the problem where the students and teacher cooperate. Teachers have a tendency to take all mathematical activity as an object of study and of teaching, which often leads them to intervene and replace an initial mathematical situation that would have permitted an authentic activity on the part of the students by a study of the mathematical circumstances and a lesson about that. We call the replacement of a situation (in particular a mathematical one) by one of its meta-situations a *metadidactical slippage*. Such a slippage tends to replace a *connaissance* by a *savoir*, which can be more easily monitored, and is assumed to be more certain (for example an algorithm), more general (a principle, a heuristic), or of a superior order (an axiom, a theorem). Alternatively, it may simply enrich the environment of a *connaissance* (its conceptual map) by examples and analogies.

This replacement may or may not be favorable, depending on the circumstances. We must therefore distinguish between the slippages that make it possible to return to the initial situation and solve it and those that do not. By a recursive process, the latter may open up a chain of uncontrolled successive metadidactical slippages that form a serious digression.

Clearly, meta-slippage is not restricted to didactical situations. It is even one of the fundamental elements of the construction of *connaissances* and the organization of *savoirs*.[1] On the other hand, within education, the lack of understanding of this slippage causes serious difficulties, sometimes within the classes themselves, depending on the culture of the teachers, but especially in macro-social didactical processes. Misjudgments and erroneous decisions follow from a lack of consciousness and understanding of the didactical nature of the phenomenon.

Various Examples of Uncontrolled Chains of Metadidactical Slippages

A recent and spectacular example was given by the representation of logic using set theory, then of set theory by naïve set theory, then of naïve set theory by Venn diagrams, then of Venn diagrams by function graphs mapping set to set, with a whole array of nomenclature, conventions and properties. Another example is that of substituting heuristics based on the works of Polya for the teaching of problem solving itself.

But all the chapters of mathematics are sprinkled with traces of this phenomenon. For example, the arithmetic resolution of linear problems to adapt to various constraints generated, in the course of history, very rich vocabularies, methods and algorithms. Their difficulty and variety resulted in a collection of mnemonic techniques like cross multiplication and topic-specific algorithms that let students apply the methods without understanding them.

Periodically swept out of mathematics by inventions – like algebra – they persist and unfortunately continue to increase and to weigh down teaching. They are imposed by the pressure of the social milieus that consider them to be cultural treasures necessary for civilized mathematics.

The Slippages Studied in This Curriculum

The present curriculum was conceived in the early 1970s, in perfect innocence with respect to this concept of slippage, but with attention paid to the relationships between that which was taught, the *savoirs*, and the means of understanding and learning it: the *connaissances*. At the time, they were treated in contrasting pairs: non-verbal decisions and schemas; languages and formulations; proofs and repertoires of validation. Our attention was swiftly drawn to the control of recursive faults and how they led off-course. The use of function graphs had already been removed from the national programme, that of representations by arrows had been

---

[1]Piaget's theory of "equilibration" maintains that all learning occurs by chains of (not necessarily didactical) meta-situations.

confined without derogation to diagrams (ratios of natural numbers and functions). The students imitated their teachers, who used these arrows without comment or specific lessons. The use of arrows was never the object of any course, or discourse, or testing, only, occasionally, of corrections. Using them was a private decision, the expression was expected to remain mathematical or "concrete". We were able to demonstrate that it was possible to use the arrows as rhetorical arguments on condition that they not be made objects of teaching. But we were also able to observe that it was impossible to produce this usage among the teachers and to avoid having the arrows give rise to metadidactical slippages, because normal teaching practices had established firm habits in the teachers. This research and others led us to examine the implicit didactical contract and its paradoxes.

It was only after discerning this phenomenon that we introduced the lessons we called "supplementary sessions", set up in two ways for the study of metadidactical slippage: the introduction of the classification of similar problems and a use of arrows between points of the plane (thus not in a diagram of the sort previously used.) Along with direct objectives that were very positive for the students, these lessons presented some dangers, but they were intended to permit us to study experimentally the conditions of a specific resistance of teachers to the sequence of slippages. As we have stated elsewhere, circumstances unfortunately did not permit us to pursue this study as we would have liked.

Evaluations

We have pointed out repeatedly the gap that exists between different views of what constitutes the results of a lesson: on the one hand, something that makes it possible to tackle situations and knowledge that would have been impossible to tackle before the lesson, and on the other something that can be evaluated by way of students' answers to precise questions, standardized or not. This difference can clearly be explained by the *connaissances* which, by definition, are not easily exportable outside of their original situation. We frequently made the experiment of comparing the answers of students to the same questions according to the situation in which the questions were posed, and we established – as is well-known – the importance of the conditions of the questioning: the status of the person asking the question, the formulation, etc.

Assessments – formal or not – were very dense. The teachers gave their opinions and proposed decisions that seemed to them to be required to improve the results. We only retained situations where successful results could be obtained without appreciably more effort than was needed for success in regular classrooms. We were allowed to make use of anonymous evaluations provided by the inspectors, and we responded immediately if results were poor.

Evaluations were considered to apply to teaching situations proposed or realized and not to teachers or even individual students. We did, however, occasionally take students in difficulties out for special sessions (as we did for Gaël, for example (Brousseau & Warfield, 1999)).

These studies allowed us to make, in 1978, a prognosis about the effects of abusive use of unconsidered diffusion of information about the school system. Not only the results of automatic massive evaluations, but also work on the schools conducted in inappropriate scientific domains to prove the legitimacy of their inferences in the didactical domain, reached populations who had every right to be interested, but used the information to reach ill-founded conclusions.

We denounced as consequences the concentration on *savoirs* to the detriment of *connaissances*, the underestimation of the capabilities of the students, the prolongation of studying time, the mincing up and multiplication of secondary didactical objectives, the individualization of teaching, and the use of tests as means of learning and then of teaching and then as an object of teaching and finally as a representation of knowledge itself.

Bibliography

Broin, D. (2002). *Arithmétique et algèbre élémentaires scolaires*. Unpublished Thèse Université Bordeaux 1.

Brousseau, G. (1965). *Les mathématiques du cours préparatoire. Paris:* Dunod.

Brousseau, G. (1970). *Processus de mathématisation, Exemple de processus de mathématisation: l'addition dans les naturels: CP-CE1*, La mathématique à l'école élémentaire (pp. 428–442, 442–457). Paris: APMEP (1972).

Brousseau, G. (1980). *Problèmes d'enseignement des décimaux* (Recherche en didactique des mathématiques, Vol. 1.1, pp. 11–59). Grenoble: La pensée sauvage.

Brousseau, G. (1981). *Problèmes de didactique des décimaux* (Recherches en didactique des mathématiques, Vol. 2.1). Grenoble: La pensée sauvage.

Brousseau, G. (1983). Les obstacles épistémologiques et les problèmes en mathématiques. In *Recherches en Didactique des Mathématiques.* Vol 4, n°2, (pp. 165–198). Grenoble: La pensée sauvage.

Brousseau, G. (1995). *Notes à propos de l'article de Thurston. On Proof and Progress in mathematics.* http://guy-brousseau.com/2100/notes-a-propos-de-larticle-de-thurston-on-proof-and-progress-in-mathematics-1995/

Brousseau, G. (1997). *Theory of didactical situations in mathematics* (N. Balacheff, M. Cooper, R. Sutherland, & V. Warfield, Trans.). Dordrecht: Kluwer Press.

Brousseau, G. (1998). *La théorie des situations didactiques. Recueil de textes de Didactique des mathématiques 1970–1990* présentés par M. Cooper, N. Balacheff, R. Sutherland, V. Warfield. Grenoble: La pensée sauvage.

Brousseau, Guy & Antibi, André (2000). La dé-transposition des connaissances scolaires. In *Recherches en Didactique des Mathématiques* Vol. 20/1, (pp. 7–40). Grenoble: La pensée sauvage.

Brousseau, N., & Brousseau, G. (1987). *Rationnels et décimaux dans la scolarité obligatoire.* de Bordeaux: IREM.

Brousseau G. & Warfield V.M. (1999). The case of Gaël, *Journal of Mathematical Behavior,* 18(1), 7–52(46). Elsevier.

Brousseau, G., Brousseau, N., & Warfield, V. (2002). An experiment on the teaching of statistics and probability. *Journal of Mathematical Behavior, 20,* 363–441. Elsevier.

Brousseau, G., Brousseau, N., & Warfield, V. (2004). Rationals and decimals as required in the school curriculum Part 1: Rationals as measurement. *Journal of Mathematical Behavior, 23,* 1–20. Elsevier.

Brousseau, G., Brousseau, N., & Warfield, V. (2007). Rationals and decimals as required in the school curriculum Part 2: From rationals to decimals. *Journal of Mathematical Behavior, 26,* 281–300. Elsevier.

Brousseau, G., Brousseau, N., & Warfield, V. (2008). Rationals and decimals as required in the school curriculum. Part 3: Rationals and decimals as linear functions. *Journal of Mathematical Behavior, 27*, 153–176. Elsevier.

Brousseau, G., Brousseau, N., & Warfield, V. (2009). Rationals and decimals as required in the school curriculum. Part 4: Problem-solving, Composed Mapping and Division. *Journal of Mathematical Behavior, 28*, 79–118. Elsevier.

Flanders, Ned A. (1976). Analyse de l'interaction et formation. In Morrison, A. & Mc Intyre (eds) *Psychologie sociale de l'enseignement.* Paris: Dunod.

Gonseth, F. (1974). *Les mathématiques et la réalité*, (1936). Paris: Librairie Albert Blanchard.

Gras R., et Kuntz P. (2008). An overview of the Statistical Implicative, Statistical Implicative Analysis, R. Gras, E. Suzuki, F. Guillet and F. Spagnolo, Eds, Springer-Verlag, Berlin-Heidelberg, p. 11–40.

Lebesgue, H. (1975). *La mesure des grandeurs*, (1931–1935). Paris: Librairie Albert Blanchard.

Odier, Antoine. (1955). *Souvenirs d'une vielle tige.* Arthéme Fayard.

Ratsimba-Rajohn, H. (1982). *Eléments d'étude de deux méthodes de mesures rationnelles* (Recherches en didactique des mathématiques, Vol. 3.1). Grenoble: La pensée sauvage.

Rouchier, A., et al. (1980). *Situations et processus didactiques dans l'étude des nombres rationnels positifs* (Recherches en didactique des mathématiques, Vol. 1.2). Grenoble: La pensée sauvage.

Thurston, W. P. (1994). On proof and progress in mathematics. *Bulletin of the American Mathematical Society, 30*, 161–177, with commentary on pp. 178–207.

Warfield, Virginia McShane. (2007). *Invitation to Didactique.* USA: Xlibris.

Whitney, H. (1968). The mathematics of physical quantities, Part I Mathematical model for measurement. *The American Mathematical Monthly, 75*, 115–138.

Whitney, H. (1968) . The mathematics of physical quantities, Part II Quantity, structure and dimensional analysis. *The American Mathematical Monthly, 76*, 227–256.

Index

A

Addition, 12, 18–20, 43, 48, 63, 74, 159
A-didactical, 147, 150–153, 157
A-didactical situations, 148, 149, 201
Algorithms, 19, 65, 78, 84, 92, 100, 125, 131, 148, 155, 206
Analogy(ies), 88, 205
Approximation, 23, 26, 49, 65, 162
Arrows, 85, 99, 118, 120, 121, 206, 207
Assessment(s), 20, 25, 130, 141, 145–147, 207

B

Balance beam, 26
Bourbaki, 167
Bracket, 33, 36–38, 40, 41, 49, 134
Brossard, M., 189

C

Cabri geometry, 201
Castelnuovo, E., 170
Centre des Observations pour Recherches en Éducation en Mathématiques (COREM), 1, 6–8, 100, 127, 128, 166, 168, 175, 179, 182, 184, 197
Choquet, G.A.A., 170
Codes, 10–11, 28
Colmez, 170
Comenius, 190
Commensuration, 6, 10–15, 26, 29–31, 93, 131, 161
Communication(s), 11–12, 15, 27–30, 148, 152, 156, 199
Connaissance(s), 89, 130–133, 135, 136, 140–152, 155, 156, 163, 189, 195–197, 199–200, 203–208

Construction, 26
Construction paper, 28
Constructivist, 5, 49, 128, 129
COREM. *See* Centre des Observations pour Recherches en Éducation en Mathématiques (COREM)
Cultural knowledge, 54

D

Decimal filters, 39–41
Decimal fractions, 45, 47, 48, 78
Decroly, O., 169
Devolution, 149, 151
Dewey, J., 169
Didactical, 2, 7, 8, 18, 20, 23, 88, 89, 92, 99, 129, 132–134, 136, 138, 146–150, 152–157, 159, 161, 165–167, 174, 178, 181, 184, 185, 187–192, 194–197, 200–201, 204, 208
contract, 197, 202–203
situation, 89, 200–201
Distributivity, 74
Division, 24–25, 42, 48–50, 54, 55, 58, 59, 85, 89–100, 112, 121–125, 163, 178, 204
Drill, 20, 47, 59
Dunod, 191
Dunot, 170

E

Eiffel, 180
Epistemological obstacles, 203–204
Equivalence, 6, 16–17, 30, 93, 95
Equivalent, 15–16, 22, 40, 99
Evaluation(s), 25, 36, 121, 132, 149–150, 154–157, 184, 207–208

F
Félix, L., 170
Freinet, C., 169

G
Gattegno, C., 170
Glaeser, G., 181
Glasses, 6, 26, 27
Gonseth, F., 190
Gréco, P., 170, 185

H
Hennequin, P.L., 185
Heuristics, 88

I
Implicit models, 196–197, 200, 201
Institutes for Research in Mathematics
    Education (IREM), 4, 7, 179–183,
    185, 186
Instituteur, 188
Institutionalization, 2, 138, 140–142, 148,
    201, 202
Institutionalized, 49, 50, 57, 65, 77, 82, 96,
    120, 132, 151
IREM. *See* Institutes for Research in
    Mathematics Education (IREM)

L
La Borderie, R., 170
Lichnerowicz, 170–171, 175, 176, 187–197
Linear, 51, 70, 78–89, 93, 94, 107–110, 114,
    115, 117, 119, 121–123, 158, 159
Linearity, 70, 106, 158
Linear mappings, 51, 79, 81, 93, 94, 107, 110,
    114, 115, 117, 121–123, 160, 161

M
Malgrange, B., 185
Mathematical situations, 132, 196–197, 200,
    202, 205
Mathematical slippages, 206
Mathematization, 86, 189, 196
Mental calculations, 19, 21, 61
Metadidactical slippage, 88, 89, 99, 121
Metaphor(s), 30, 91, 100, 136, 193, 205

Meta-situation, 205
Milieu, 138, 147, 191, 200, 201, 205
Multiplication, 18, 24, 30, 48, 57, 59, 62, 74,
    75, 78, 86, 90, 94–98, 100, 110, 113,
    159, 163, 178

N
Number line, 41–45, 139

O
Obsolescence, 155–156
Odier, A., 180
Operator, 56
Optimist, 63–65, 67, 69, 78, 97, 98, 100,
    136, 163
Ordered pairs, 15–17, 29, 97, 161

P
Papy, G., 169, 170
Partitioning, 28, 92, 93
Pedagogy, 170, 171, 187, 188
Pescarini A., 170
Piaget, J., 169, 172, 189–191, 193
Pisot, C., 169
Pluvinage, F., 181
Polya, G., 88
Precision, 2, 15, 26, 28, 86, 91, 142, 144, 153,
    155, 159
Proportional, 63, 64, 69, 70, 72, 73
Puzzle, 6, 51–55, 61, 69, 70, 136, 159

R
Ratio, 63, 66, 68, 81, 82, 86, 93, 119
Reciprocal, 73, 94–96, 98, 118–123, 160
Representation, 41–45

S
Savoirs, 89, 130 132, 135, 137, 139 145,
    147–150, 152, 153, 155, 156, 163,
    189, 191, 192, 195–196, 199–200, 203,
    205, 206
Servais W., 170
Situations
    of actions, 140, 152–153, 192, 201, 202
    of communication, 68, 140, 152, 192
    of evaluation, 149–150
    of formulation, 140, 151–153, 201, 202

of institutionalization, 139, 148–149,
    192, 196
of validation, 140, 151, 192, 201, 202
Spiral method, 174, 176
Strategies, 12, 17, 19, 21, 34, 38, 40, 42, 52,
    53, 60, 61, 64–66, 76, 92, 139, 151, 153
Sum, 18–21, 36, 70, 81, 85, 90, 91, 94, 159

T
Tessellation, 60, 137
Touanet, H., 185
Tournament of problems, 79

U
Unsaid, 2, 157

V
Vergnaud, G., 183
Vocabulary, 12, 16, 30, 100, 120, 152, 161,
    171, 175, 178
Vygotsky, L.S., 169

Z
Zamanski, M., 169